Citizen Participation in
Global Environmental Governance

Citizen Participation in Global Environmental Governance

Edited by

Mikko Rask, Richard Worthington, Minna Lammi

earthscan
from Routledge

First published 2012
by Earthscan
2 Park Square, Milton Park, Abingdon, Oxon OX14 4RN

Simultaneously published in the USA and Canada
by Earthscan
711 Third Avenue, New York, NY 10017

Earthscan is an imprint of the Taylor & Francis Group, an informa business

British Library Cataloguing in Publication Data
A catalogue record for this book is available from the British Library

Library of Congress Cataloging in Publication Data
A catalog record for this book has been requested

ISBN: 978-1-84971-378-8 (hbk)
ISBN: 978-1-84971-379-5 (pbk)

Typeset in Times
by JS Typesetting Ltd, Porthcawl, Mid Glamorgan

Contents

List of Figures, Tables and Boxes

Figures

Tables

Boxes

List of Contributors

Annika Agger, PhD, is associate professor of Citizen Participation in the Department of Society and Globalisation, Roskilde University, Denmark. She is currently involved in research projects on new forms of local democracy and public deliberation in the context of urban and environmental planning.

Edward Andersson is the deputy director of Involve (UK) and an expert in participatory methods. He has advised a number of organizations on public engagement strategies, including the European Commission and the OECD. He is a certified facilitator, and studied citizen engagement strategies with complex multilevel technical and political debates.

Alison Atherton is a research principal at the Institute for Sustainable Futures (ISF) at the University of Technology, Sydney. She has a background in social sciences and worked with leading Australian organizations on sustainability policies and environmental management systems. At ISF she has worked on several projects related to sustainability strategies and future scenarios. She managed the Australian WWViews process.

Ravtosh Bal is a PhD candidate in the joint PhD program in public policy at Georgia Institute of Technology and Georgia State University. Her research interests are participatory technology assessment and public participation in policy-making. She holds MA and MPhil degrees in sociology from the University of Delhi and has worked in the Government of India as a member of the central civil service.

Ulrike Bechtold is a doctor in human ecology and researcher at the Institute of Technology Assessment of the Austrian Academy of Sciences. Her research interests are in sustainability policies and participative approaches in the study of interaction between technology, human beings and natural resources. She is also a lecturer at the University of Vienna.

Bjørn Bedsted has been working as a project manager with the Danish Board of Technology (DBT) since 2004 after finishing his MA in social anthropology. At DBT he has made technology assessments in various fields, such as IT security, GM, the patent system and climate issues, often using a variety of citizen participation methods developed at DBT. He was global coordinator of the WWViews project.

Gwendolyn Blue is an assistant professor in the Department of Geography at the University of Calgary. She also holds an appointment in the Department of Communication and Culture. Her research examines transformations in governance and the politics of knowledge, particularly as these are taken up in relation to food, water and climate change.

Isabel Bortagaray, PhD in public policy, is currently working at the International Development Research Center (Canada) in the Regional Office for Latin America and the Caribbean. Her work has concentrated on the analysis of science, technology and innovation policies in the context of developing countries. This study, however, took place when she was an adjunct professor at the University Research Council, University of the Republic, Uruguay.

Nadine Brachatzek is a member of ZIRN at Stuttgart University, Germany. She has a master's degree in sociology, linguistics and mathematics from Stuttgart (2009). Her research interests are in energy consumption, future trends and simulation. She also has experience with deliberative projects and science communication.

Karetta Crooks Charles is the communications and advocacy officer at the Saint Lucia National Trust (SLNT). A Jamaican national with her first degree in media and communication from the University of the West Indies (Mona), she worked as a teacher in Colombia and a radio news reporter in Saint Lucia prior to assuming her current position. One of her programmes at SLNT is the Youth Environment Forum, which empowers youths aged 7–18 to become advocates for environmental and heritage conservation.

Jason Delborne is an assistant professor of Liberal Arts and International Studies at the Colorado School of Mines. He has studied highly politicized scientific and technological controversies. In 2010 he co-edited *Controversies in Science and Technology: From Evolution to Energy* (Mary Ann Liebert, Inc.) and was awarded the David Edge Prize for the best article in the area of science and technology studies by the Society for Social Studies of Science.

Edna Einsiedel is a professor of Communication Studies, University of Calgary. She served as a project director for the Canadian WWViews consultation. Her research has examined the social contexts of new technologies and the roles and participation of publics in the governance of science-based policy questions.

Rüdiger Goldschmidt is a graduate sociologist interested in empirical research and methodology related to citizen participation, risk perception and corporate social responsibility. Since 1998 he has worked in the field of evaluation, including for example the 'Meeting of Minds' and the 'European Citizens' Consultations' projects. In his

doctoral dissertation study he focuses on the criteria necessary for evaluating deliberative processes.

Søren Gram, PhD, is a cultural sociologist and works as a senior project manager at
the Danish Board of Technology (DBT). Before joining DBT in 1997 he worked as a
researcher at the University of Copenhagen. At DBT he has worked with most of the
technology assessment methods involving different stakeholders. He was a member of
the WWViews coordination team and project leader of the Danish WWViews citizen
meeting.

Leonhard Hennen, doctor in sociology from Technical University Aachen, has extensive experience of technology assessment projects at the Office of Technology
Assessment at the German Parliament and Karlsruhe Institute of Technology. Since
2006 he has been the coordinator of the European Technology Assessment Group
(ETAG), a consortium of eight European TA institutes carrying out projects on behalf
of the European Parliament.

Jade Herriman is a research principal at the Institute for Sustainable Futures (ISF)
at the University of Technology, Sydney. She has a background in ecology and local
government, and has worked in several community engagement and participatory
democracy projects. She is particularly interested in community-scale responses to
sustainability challenges and individual, social and organizational processes related to
environmental policies.

Birgit Jæger is a professor of Science, Technology and Society at the Department
of Society and Globalisation, Roskilde University, Denmark. She has studied the
development of multimedia in local and international contexts, the role of eGovernment
in Denmark and the use of ICT by senior citizens. Currently she is working on the
research project 'Collaborative Innovation in the Public Sector', funded by the Danish
Research Council

Erling Jelsøe is an associate professor at the Department of Environmental, Social and
Spatial Change, Roskilde University, Denmark. His research includes social studies of
food production, food policy and sustainable food production, public perceptions and
regulation of new biotechnologies and public participation in science and technology.

Lars Klüver has been the director of the Danish Board of Technology (DBT), the
parliamentary technology assessment institution of Denmark, since 1996. He has been
active in the development of public participation methodology and the use of such
methods for informing and advising policy-making since 1986. He was the initiator of
WWViews and president of the WWViews Alliance. He has an MSc in ecology and
environmental biology.

Martin Knapp, PhD in biology, is a senior researcher at the Institute for Technology Assessment and Systems Analysis (ITAS) at Karlsruhe Institute of Technology, Germany. He has experience of deliberative methods, e.g. scenario-building workshops, used in projects focused on environmental topics, such as agricultural biotechnology, climate change and sustainable land use. He was the coordinator of the German WWViews.

Maarit Laihonen is a doctoral student at Aalto University School of Economics, Helsinki, Finland and a master's student in social and moral philosophy. She was one of the 17 Finnish climate advocates in the British Council's Challenge Europe project in 2010–2011. She supported the Finnish WWViews project as a research trainee.

Minna Lammi, PhD in political science from the University of Helsinki, is a head of research at the National Consumer Research Centre, Finland. Her research focuses on the construction of consumer society and the interconnection between consumerism and citizenship. She has published in the *International Journal of Consumer Studies* among others and has edited several books.

Marila Lázaro, PhD in philosophy, science and society (Basque Country University), is associate professor of the Unit of Science and Development at the School of Science in Montevideo (Universidad de la República, Uruguay). She is a founding member of Simurg, an NGO aimed at fostering social appropriation of scientific knowledge through public participation.

Jennifer Medlock is a PhD candidate in the Department of Communication and Culture at the University of Calgary, Canada. Her research focuses on initiatives to encourage public engagement with scientific and technological policy.

Michael Ornetzeder is a senior researcher at the Institute of Technology Assessment at the Austrian Academy of Sciences and a lecturer at the University of Natural Resources and Applied Life Sciences in Vienna. He focuses on science and technology studies, participatory technology assessment, user innovation, social learning and innovation networks. Currently he studies the transition of the energy system and climate change issues.

Louise Phillips (BA Hons, Strathclyde; MSc, PhD, LSE) is associate professor in Communication Studies at the Department of Communication, Business and Information Technologies, Roskilde University, Denmark. Her research interests include the production and communication of knowledge and participatory, dialogue-based approaches to communication theory and practice.

Jacqueline Pomeroy received her bachelor's degree in politics and public policy analysis at Pomona College in Claremont, California, and was a research assistant for this volume. She wrote her senior thesis on the portfolio management model of school reform as implemented in the Los Angeles Unified School District. Her thesis focused on how stakeholder groups organized and garnered community support for school models. She is currently working in Los Angeles as a director of a grass-roots campaign in the 2012 election cycle.

Christiane Quendt works as a junior researcher at the Institute for Technology Assessment and Systems Analysis (ITAS) at the Karlsruhe Institute of Technology, Germany. She has master's degrees in communication and media science, english studies and economics. She is interested in different participatory and deliberative methods, and worked recently in projects related to nanotechnology in society and media.

Mikko Rask, PhD in environmental strategies and technology assessment, is a senior researcher at the National Consumer Research Centre, Finland. He has experience of several national and EU projects on technology assessment, foresight and sustainability policy. He has published widely on these themes and lectured on science and technology and futures studies in several universities. He was the coordinator of WWViews Finland.

Ortwin Renn is a professor and chair of Environmental Sociology and Technology Assessment at Stuttgart University, Germany. He directs the Interdisciplinary Research Unit for Risk Governance and Sustainable Technology Development (ZIRN) and the non-profit research institute DIALOGIK. He is interested in risk governance, political participation and technology assessment, and has published over 30 books and 250 articles, including *Risk Governance* (Earthscan, 2008).

Petteri Repo, PhD in economics, is head of research at the National Consumer Research Centre, Finland and adjunct professor (docent) at the Aalto University School of Economics. His research interests include user participation in product development and policy issues relating to consumption.

Jen Schneider is an assistant professor of Liberal Arts and International Studies at the Colorado School of Mines in Golden, Colorado. She is focused on media and communication studies, and published and taught on topics related to energy policy, science and risk communication, public engagement and engineering. She recently co-edited *Engineering and Sustainable Community Development* (Morgan and Claypool, 2010).

Thea Shahrokh is a researcher at Involve, the UK's leading public participation research organization. She has studied innovative methods of public engagement, especially in the development of healthcare and information and communication technology systems. She was the project manager of WWViews UK, and has recently worked hard to raise Involve's profile in the field of international deliberation.

Mahshid Sotoudeh, doctor in sustainable development, is a senior researcher at the Institute of Technology Assessment of the Austrian Academy of Sciences and an associate professor of Technology Assessment and Sustainable Development at the University of Technology in Graz. She has studied national and international strategies on sustainable engineering education. In 2009 she published *Technical Education for Sustainability: An Analysis of Needs in the 21st Century* (Peter Lang).

Päivi Timonen, PhD in consumer economics, is the research director of the National Consumer Research Centre, Finland. She has studied consumers' everyday heuristics in dealing with environmental information. Her research interests are in the area of new markets and user needs. She has published in the *European Journal of Innovation Management* and the *International Journal of Technology and Human Interaction*, among others.

Ana Vasquez Herrera, MSc in science communication, is currently working at the Faculty of Science, University of the Republic, in Montevideo on public participation in science and technology (S&T) and in the public sector as part of the S&T Education and Culture Office. She is a founding member of Simurg, an NGO aimed at fostering social appropriation of scientific knowledge through public participation.

Stuart White is the director of the Institute for Sustainable Futures (ISF) at the University of Technology, Sydney. Professor White has directed a number of participatory projects, including the world's first combined citizen jury and televote in 2000 and the Yarra River Values Forum in 2006. His research focuses on improved decision-making for public policy outcomes in resource use and infrastructure. He has published widely on sustainable futures and is a regular commentator on sustainability issues in the Australian media.

Richard Worthington is a professor of Politics at Pomona College, USA, where he chairs the Program in Public Policy Analysis. He chairs the board of trustees of the Loka Institute, which advocates democratic approaches to scientific and technological policies. He has published widely on environmental activism, science policy and politics in the Americas, including *Rethinking Globalization: Production, Politics, Actions* (Peter Lang, 2000). He coordinated the five WWViews sites in the USA.

Preface

This book project started in March 2009, when the idea of collecting cross-cultural lessons from the first-ever global citizen deliberation process, World Wide Views on Global Warming (WWViews), was introduced to a group of researchers who participated in a training seminar in Copenhagen organized by the Danish Board of Technology (DBT). A call for papers was opened immediately after the WWViews deliberation process in September 2009, and with the help of external peer reviewers, the editors organized the various research contributions to this volume as a multidimensional case study about citizen participation in global environmental governance. DBT was the initiator of WWViews, and this book project could not have taken place without their support and flexibility regarding our efforts to integrate research and evaluation activities in the ongoing deliberation project. We are especially grateful to Bjørn Bedsted, Lars Klüver and Søren Gram from DBT, who also agreed to tell their own 'story of WWViews' in this volume.

The research contributions to this volume are written by some four dozen researchers and practitioners who were involved in organizing and/or studying the national WWViews deliberations. We are grateful to the authors for their high-quality contributions and commitment to the book project, completed on a voluntary basis. We also appreciate the contributions by the following peer reviewers of this volume: Christina Benighaus, Trevor Campbell, Guillermo Foladori, Leena Haanpää, Katri Huutoniemi, Alan Irwin, Stefanie Jenssen, Leena Jokinen, Sirkku Juhola, Juha Kaskinen, Jenni Kauppila, Eeva Kuntsi-Reunanen, Richard Langlais, Jane Lehr, Sylvia Lorek, Johanna Mäkelä, Juri Mykkänen, Mari Niva, Gwen Ottinger, Taru Peltola, Sinikka Pesonen, Roli Varma and Marja Vehviläinen – as well as the four anonymous book reviewers commissioned by the publisher.

During the book project two international meetings were organized in 2010, a research workshop in Snekkersten, Denmark and a 4S (Society for Social Studies of Science) conference session in Tokyo, where many of the authors presented their manuscripts and other researchers interested in deliberative democracy gave their feedback and comments. We are especially thankful to Nicolas Baya Laffite, Simon Burrall, Netra Chhetri, Susan Cozzens and John Dryzek for their constructive ideas and viewpoints.

The editorial process has been laborious, and we are grateful for all the help that we have received from our colleagues and organizations. We especially thank

Jacqueline Pomeroy for her invaluable help in commenting and proofreading many of the manuscripts, Terttu Sarpiola for taking care of systematizing the references and Taina Pohjoisaho for strengthening the visualizations in the introductory and conclusion chapters. We are also grateful to Charles Balter, Dawn Bickett, Trevor Bisset, Eva Heiskanen, Mikael Johnson, Hilary LaConte, Mika Pantzar, Petteri Repo, Mika Saastamoinen, Päivi Timonen and Grace Vermeer, who have provided their comments and collegial support during the book process. We thankfully acknowledge the professional editorial support of Alison Kuznets and Anna Rice of Earthscan.

National Consumer Research Centre (NCRC) Finland has provided personnel and financial resources in coordinating the book project: we thank Director Eila Kilpiö for the generous support that she has provided to this research effort. The US contributors were supported by National Science Foundation Award #0925043. We are also grateful to Director Robby Berloznik from the Flemish Institute Society and Technology (IST), who institutionally supported this book project, as well as Loka Institute (USA) and the Institute for Sustainable Futures (Australia).

Mikko Rask, Richard Worthington and Minna Lammi
Helsinki, Finland and Claremont, California, USA, April 2011

List of Acronyms and Abbreviations

ACF	advocacy coalition framework
AOSIS	Alliance of Small Island States
ASTC	Association of Science and Technology Centers
CARICOM	Caribbean Community
CBR	community-based research
CC*	consensus conference
COP15	15th meeting of the Conference of Parties to the United Nations Framework Convention on Climate Change
CSM	Centre for Social Markets (Bangalore, India)
CSSP	Centre for Studies in Science Policy (New Delhi, India)
DBT	Danish Board of Technology (Denmark)
DD	deliberative democracy
DDP	deliberative democratic process
ECAST	Expert and Citizen Assessment of Science and Technology
ECC	European Citizens' Consultation (2007 and 2009)
ETAG	European Technology Assessment Group
GDP	gross domestic product
GHG	greenhouse gas
GMO	genetically modified organism
ICT	information and communication technologies
IDS	Institute of Development Studies (UK)
IIED	International Institute for Environment and Development
IPCC	Intergovernmental Panel on Climate Change
ISF	Institute for Sustainable Futures (Australia)
IST	Flemish Institute Society and Technology
ITA	Institute of Technology Assessment (Austria)
ITAS	Institute for Technology Assessment and Systems Analysis (Germany)
NBSAP	National Biodiversity Strategy and Action Plan (India)
NCRC	National Consumer Research Centre (Finland)
NGO	non-governmental organization
OECS	Organization of Eastern Caribbean States
OTA	Office of Technology Assessment (USA)
PEP-NET	Pan European eParticipation Network

pTA	participatory technology assessment
SDES	Sustainable Development and Environment Section (Saint Lucia)
SEP	Sustainable Energy Plan (Saint Lucia)
SIDS	small island developing states
SLNT	Saint Lucia National Trust
S&T	science and technology
STS	science, technology and society
TA	technology assessment
TAMI	Technology Assessment Methods and Impacts
UNFCCC	United Nations Framework Convention on Climate Change
WWViews	World Wide Views on Global Warming
ZIRN	Interdisciplinary Research Unit on Risk Governance and Sustainable Technology Development

Part I
Introduction

1

Towards a New Concept of Global Governance

Mikko Rask and Richard Worthington

On 26 September 2009, ordinary citizens convened at 44 sites in 38 countries to discuss the issues that would be on the agenda at the December 2009 United Nations Framework Convention on Climate Change (UNFCCC) summit in Copenhagen, also known as COP15.[1] Armed with 40-page briefing reports they had received before the event and seated at tables of six to eight (approximately 90 people in total at each site), participants debated and voted on a common set of policy choices and developed their own recommendations to policy-makers in the first global citizen consultation in history. In the words of the project organizer, The Danish Board of Technology (DBT), World Wide Views on Global Warming (WWViews) aimed

> *... to give a broad sample of citizens from across the Earth the opportunity to influence global climate policy. An overarching purpose was to set a groundbreaking precedent by demonstrating that political decision-making processes on a global scale benefit when everyday people participate.* (www.wwviews.org/node/223)

This book examines what happened on 26 September and after, as well as the opportunities for deliberative global governance[2] that lie ahead. The first sentence of DBT's aspirations for WWViews quoted above suggests both the richness and the challenge of our task. The project clearly gave an unprecedented range of citizens worldwide a unique and meaningful opportunity to deliberate on global climate policy, but its actual and potential influence is hard to conceptualize, and harder yet to measure. Nonetheless, the chapters in this volume generally support the claims that (1) the deliberations provided valuable insight into the informed and considered views of ordinary people,[3] which otherwise are virtually absent from global policy venues; but that (2) the impacts on climate policy were very limited. More importantly, the contributors to this volume engage the empirical record of this unprecedented global event at ground level, triangulate it with more than two decades of experience with deliberative democracy (DD) and participatory technology assessment (pTA), and develop nuanced accounts of the outcomes that advance both scholarly and practical discussions of deliberative global governance.

The participatory emphasis that is at the core of WWViews extends to the method followed by most contributors to this volume, who can be characterized as action researchers. The authors include university professors as well as practitioners of deliberative democracy, operating in an action research mode, where the producers of a social intervention (WWViews) also study it. This has many of the qualities of an experiment, but differs because an intervention aims to change policy and the complex set of social forces connected to it. WWViews also bears some similarities to participatory action research or community-based research (CBR), most obviously that it is participatory at its core and pervaded with an ethos of incorporating the voices of ordinary people into the policy-making system. However, WWViews differs from CBR in the critical sense that the latter refers to those unique situations in which the people experiencing a problem play a central role in producing knowledge about it. While everyone in some sense 'experiences' global warming, the provenance of this project resides with an expert organization, DBT, and the members of the World Wide Views Alliance that they enrolled in the project. While the process was highly participatory – e.g. project managers worldwide made extensive contributions to forming the questions that were put to citizens on 26 September – the ordinary people who ultimately deliberated on climate change had virtually no role in conceiving or controlling the project. This signals an important distinction for global problems, which seem to rely more heavily than community initiatives on expert intermediaries who, among other things, can help participants navigate the many complexities of multiple levels of governance and problem-solving.

Our central concern in this volume is the value of global citizen deliberation. Seeing little prospect for direct measures of such value, and appreciating the multiplicity of perspectives embedded in the matter, the concern is addressed through subsidiary questions such as: What hopes and aspirations inspire hundreds of professionals to voluntarily venture into such a process? What are the motivations, expectations and experiences of the thousands of citizens who voluntarily spend their time in considering climate policy issues? What are the perspectives of policy-makers and researchers on the role of citizen deliberation at different levels of policy-making? How did WWViews impact climate policy and other political processes? What is the methodology of global citizen deliberation that works in diverse cultural contexts, and how could it be further developed in the future? If the reader has a better sense of WWViews' value after examining the following chapters, where these and other questions are addressed, we will count the exercise as a success.

Like the WWViews project itself, this book results from the voluntary activity of people across countries and continents, who started to work together and collect ideas and lessons on how to develop global citizen deliberation on the basis of the experience from WWViews. Some 40 researchers submitted abstracts in response to the call, proposing different analytical focuses on the study of WWViews in different cultural contexts. During the editorial process, 12 research articles were finally completed and included as chapters in this book. To ensure scientific quality, all research articles were submitted to a double-blind peer review.

In this introductory chapter, we provide background information about the WWViews project and address the role of research within it. Since WWViews was the first global citizen deliberation in history, we also aim to identify and explore the kinds of issues that emerge from such a novel endeavour, i.e. issues that researchers find interesting, characteristic of this particular deliberation exercise, and worth further consideration. We also give a word to the originators of the WWViews concept, Bjørn Bedsted, Søren Gram and Lars Klüver from DBT to present their ideas about the project and its organization.

The research chapters are organized into four parts:

- Understanding the Trend of International Deliberation – two articles studying theoretical starting points and recent examples of international deliberation projects;
- Evaluating the WWViews Process and Results – a section analysing and interpreting WWViews process and outcomes from different evaluative perspectives;
- Discussing Cultural Variation and Local Engagement of Expertise – studies reflecting the contextual particularities of WWViews sites in different countries, and reporting on more and less successful efforts to interact with local actors; and
- Studying Policy and Media Impacts – a section focused on understanding the bottom-line effects of WWViews.

Finally, the main lessons of the book are discussed in the conclusion. Given our action research orientation, we also proffer focused insights and recommendations that might have practical value for future global deliberations. The conclusion draws on an international workshop of researchers and practitioners organized in Snekkersten, Denmark in June 2010 to discuss the challenges of global deliberation processes, in addition to the chapters in this book.[4]

In the following sections of this introduction, we map issues related to global deliberation that emerge from the research articles. These include foundational questions pertaining to the theoretical roots and social driving forces of deliberative democracy; evaluations of the fairness, competence and pragmatic aspects of the deliberation process; discussions about the social and political impacts of WWViews consultations that took place in different policy cultures and contexts; as well as observations of emerging issues that need further attention and research to achieve better understanding of their meaning for the future of global deliberative democracy (see Figure 1.1 below). In the concluding chapter of this volume, we reprise these issues in order to identify elements that provide firm ground for developing future international deliberations, and questions that deserve further research.

The WWViews event and its results[5]

The WWViews event followed a uniform scheme in all partner countries. Approximately 4000 citizens from 38 countries (see Table 1.1) chosen to reflect the demographic

Table 1.1 *Number of participants per WWViews site and country*

Continent	Country and site	Participants
Asia	Bangladesh	100
	China	70
	Chinese Taipei (Taiwan)	108
	India – New Delhi	47
	India – Bangalore	30
	Indonesia – Jakarta	100
	Indonesia – Makassar	105
	Japan	105
	The Maldives	72
	Vietnam	90
Africa	Cameroon	75
	Ethiopia	98
	Egypt	109
	Malawi	104
	Mali	101
	Mozambique	97
	South Africa	100
	Uganda	105
Australia	Australia	105
Europe	Austria	96
	Belgium (Flanders)	82
	Denmark	96
	Finland	107
	France	98
	Germany	81
	Italy	90
	Netherlands	94
	Norway	95
	Russia	94
	Spain	100
	Sweden	52
	Switzerland	72
	United Kingdom	100
North America	Canada	103
	US – Arizona	87

Continent	Country and site	Participants
	US – California	63
	US – Colorado	66
	US – Georgia	41
	US – Massachusetts	78
South America	Bolivia	102
	Brazil	93
	Chile	86
	Uruguay	96
	St Lucia	67
Average per country		88
Total number of participants		3860

diversity in their respective countries and regions were provided with balanced information about climate change and COP15 negotiations, and they were given time and circumstances to deliberate with their fellow citizens.

The questions put to the citizens worldwide were designed to be of direct relevance to the COP15 negotiations. The same questions were posed in all countries in order to create comparable results that would also be meaningful when aggregated globally. Twelve questions with multiple-choice answers were chosen and clustered in four themes: climate change and its consequences, long-term climate goal and urgency, dealing with greenhouse gas emissions and the economy of technology and adaptation. To balance this restricted format, citizens were also asked to formulate and vote on their own recommendations.

Background information[6] provided to the participating citizens consisted of an information booklet and videos produced specifically for the event and aligned with the four themes around which the deliberation was structured. The 40-page booklet was sent to participants a few weeks before the event. About half of it addressed the basics of climate change science, drawing on the *Fourth Assessment Report* of the Intergovernmental Panel on Climate Change (IPCC). The rest examined the background and goals of the existing and proposed climate treaties, how to distribute responsibility for reducing greenhouse gas emissions among countries with different emission levels and economic development profiles, and how to pay for mitigating and adapting to climate change. Information videos of 5–12 minutes each that contained essential information from the booklet were played at the beginning of each thematic session. This reinforced what participants had read, and ensured a common baseline of information, given the likelihood that individual participants varied widely in both the amount of the preparatory reading completed as well as their comprehension of it. An international advisory board conducted a scientific review of the material, while its relevance and accessibility were tested by focus groups comprised of lay people in Japan, Canada, Denmark, Bolivia and the US.[7]

The design of the consultations called for approximately 100 citizens at each site to be seated at tables of five to eight people with a facilitator who had been trained to ensure that rules of respectful debate were followed. After watching the information videos, the participants engaged in moderated discussions which gave them time to listen to other opinions and reflect prior to voting. Each thematic session concluded with citizens casting their votes anonymously on two to four questions. After concluding the four thematic sessions, a final session was held in which citizens at each table wrote in their own words what they believed to be the most important recommendation to pass on to COP15 negotiators. These were posted on flip charts or taped to walls so that citizens could read all the recommendations at their site and then submit individual ballots generating a prioritized list of recommendations.

The reporting of the results took place instantaneously on the internet home page of the project (www.wwviews.org), so that anyone with internet access could – and still can – compare answers to the various questions across countries and regions, as well as political and economic groupings. The WWViews consultation started on 26 September 2009 at 9.00 am in Australia and ended 36 hours later in Arizona and California, US. In addition, three international expert panels discussed the results in facilitated debates on an internet video and blog. Some national sites, such as the Swedish city of Borlänge and the UK town of Kettering were mutually contacted through Skype meetings at the time of the consultation.

The final results of the consultation were presented about two months after the event in a policy report that was produced by the Danish coordinators, eight WWViews project managers, and one liaison officer who supported the US team and the WWViews secretariat (Bedsted and Klüver, 2009). After surveying all the results, the panel wrote that, 'one impression stands clear: The participating citizens mandate their politicians to take fast and strong action at COP15' (Bedsted and Klüver, 2009, p11).

Looking at the results from 44 sites where WWViews consultations were held, it is remarkable how consistently[8] citizens across different demographic groups and geographic regions think about the urgency of stringent climate policy. Citizens across the globe want to keep the temperature increase below 2°C; to set reduction targets not only to industrialized countries but also to fast-growing economies and low-income countries; to give priority to the creation of international financial mechanisms and technology transfer programmes; to strengthen international institutions and introduce both rewards and punishments for countries to reach their carbon emission goals. The main results of the WWViews consultation are summarized in the nine policy recommendations presented in Box 1.1.

A different picture of the thinking of citizens worldwide emerges from the 488 recommendations (from 3 to 20 per country) to policy-makers that citizens developed

---

Box 1.1 Main policy recommendations of WWViews

- *Make a deal at COP15*: fully nine out of ten urged their COP15 delegations to reach a new, binding climate change deal at COP15, rather than waiting until later.
- *Keep the temperature increase below 2°C*: almost nine out of ten viewed 2°C as a maximum acceptable goal for temperature increase. Half of the participants even want to limit the increase to the current level or return to the pre-industrial level.
- *Annex I countries should reduce emissions by 25–40 per cent or more by 2020*: the participants wanted COP15 negotiators to agree on year 2020 emissions reductions of 25–40 per cent or more below 1990 levels for Annex I countries.
- *Fast-growing economies should also reduce emissions by 2020*: WWViews participants supported the introduction of 2020 reduction targets for fast-growing economies such as Brazil, China, Chile, Egypt, India, Indonesia, South Africa and Uruguay that have substantial economic income and/or high emissions. This support is equally strong (with one exception)[9] among citizens from these countries themselves.
- *Low-income developing countries should limit emissions*: there was strong support to put in place short-term emission limits for low-income developing countries. The strongest demands came from citizens from the poorest countries themselves.
- *Give high priority to an international financial mechanism*: WWViews participants expressed a strong wish for COP15 to institute a financial mechanism that will secure funding for mitigation and adaptation in developing countries. They expected a mechanism with automatic and mandatory payments, rather than a mechanism that is subject to individual nations' voluntary contributions.
- *Punish non-complying countries*: a clear majority supported punishing countries that do not meet their commitments under a new climate deal. They also supported the introduction of incentive measures to reward behavioural change and technological development.
- *Make technology available to everyone*: participants around the world believed that it is crucial for the COP15 negotiations to give high priority to technology transfer and investments in order to meet climate targets.
- *Strengthen or supplement international institutions*: WWViews participants expect COP15 negotiators to ensure that new or stronger international institutions are put in place to advance the objectives of a new climate deal.

Source: Bedsted and Klüver (2009)

---

using their own ideas and words. These recommendations suggest the diversity of objectives, means and values that citizens find important on the way towards a world that controls its temperature in a sustainable way. Quite interestingly, along with the suggestion for solutions at the global political level, people focus attention on the role of values and educational institutions and the need to change patterns of consumption and lifestyle. In their analysis of those recommendations Lammi et al[10] (Chapter 7) argue that when lacking power in global political questions, citizens seek power from the area of 'consumerism', and call for concrete measures in those areas, such as regulating consumption and labelling.

We next turn our attention from the results of WWViews to its methodological and procedural characteristics and its political impacts. The chapters in this volume where these issues are discussed in depth draw on an empirical record that is unprecedented in studies of democratic deliberation and analyse it from a multiplicity of perspectives that are both familiar and at times novel.

Exploring the issues and characteristics of WWViews

There is a shared understanding among the contributors about the uniqueness of WWViews: citizen deliberations have occurred before, but never at a global scale. Organizing a citizen consultation in six continents and 38 countries on a single day therefore involves procedural complexity not previously encountered in the realm of deliberative democracy. What is also unique about WWViews is its aspiration to influence global policy-making through a citizen consultation process based on informed discussions.

Comparing WWViews with better-known processes and examples, by force of analogy, and applying existing theoretical and evaluative frameworks are the main approaches by which the researchers of this volume aim to explore and understand the issues emerging from WWViews. At the same time, however, there is the recognition that many of these issues are new and emerging, for which reason we include detailed empirical accounts of site-specific activities, with the intention of recording local experiences and identifying more general lessons. Both approaches are found in the following chapters, including some that adopt a mixed approach by discussing theoretical problems (e.g. the idea of representation) in the light of practical experience (e.g. issues encountered in actually forming citizen panels), leading to new insights about how the method of global deliberation could be refined (e.g. less attention to the exact number of participants can concur with a more deliberative model of public consultation).

Figure 1.1 provides a summary of the main issues discussed in the research chapters of this volume. Consensus conference (CC* as abbreviated in Figure 1.1) is a benchmark and analogy to WWViews that is used by many authors of this book. Consensus conference is another model of citizen consultation developed by DBT. In that model, citizens are invited by the Danish Parliament to learn about and discuss specific issues related to science and technology (S&T) policy. The citizens' recommendations on particular policy options are then presented to the parliamentarians, who give their response to the citizen panel, and if possible, incorporate the recommendations into Danish law. In this context, therefore, small groups of citizens function as advisers to policy-makers. The consensus conference has become a widely used and broadly studied method in the field of participatory technology assessment (pTA),[11] and therefore provides several points of reference for comparison to WWViews (see e.g. Joss, 1998; Guston, 1999; Joss and Bellucci, 2002; Kleinman and Powell, 2007; Sclove, 2000, 2010).

The application of the consensus conference process internationally, e.g. in Taiwan, South Korea, Japan and Argentina, has contributed to the hypothesis that participatory

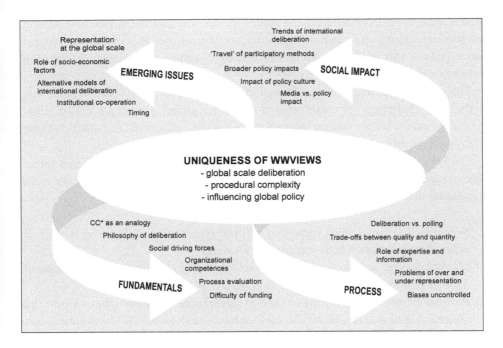

Figure 1.1 *Characteristics of WWViews and issues for global deliberative democracy*

methods 'travel well' to different policy contexts. This was also the starting point of the DBT staff, who decided to develop WWViews as a uniform approach to be consistently applied in all partner countries. That the WWViews concept actually travelled to 38 countries within a rather short time span (in approximately one year) supports the hypothesis. The chapters in this book, however, point to important refinements in the hypothesis. The modest policy impact and low recognition of WWViews by local policy-makers, stakeholders and media actors in several partner countries imply that the 'travel' too often stops at the walls of the deliberation room. This suggests that for participatory methods to travel better, more effective strategies of policy transfer and local adaptation are required. The more distant and disparate the policy context from the tradition of pTA, the greater is the need for adaptation strategies. Various means for increasing connectivity of the uniform deliberation concept to local policy context are proposed in several articles[12] including, for example, ideas for increased institutional cooperation with national policy-makers and introduction of local issues as part of the global deliberation agenda.

Methodological and procedural issues

A number of methodological and procedural issues are discussed in the following chapters. Most critical questions are related to the way that representativeness of the citizen panels is both understood and achieved; the way that communication among and between citizens and experts is organized; and the way that the deliberation process converged towards producing relevant results by combining voting and deliberation techniques.

The issue of representation is addressed in several chapters, and is undoubtedly one of the main challenges for establishing a model for global democratic deliberation. Instead of narrowly focusing on the statistical dilemma of sample construction,[13] the contributors to this volume take a more holistic account of representation.

The issues of representation are closely linked to an understanding of deliberative democracy as a model of political activity. The following questions reflect the main approaches to this issue taken by the authors:

- *Social driving forces*: what social and political forces make deliberative models of political representation both sensible and feasible?
- *Methodological solutions*: what methods are being used in other transnational deliberations and how does WWViews compare with them?
- *Rationales*: what are the political and philosophical arguments that support the organization of discussions within 'mini-publics', such as in WWViews panels, as a complement to deliberation among the entire public?
- *Limitations*: what are the practical and theoretical limitations to displacing or complementing citizens' opinion with WWViews type of panel discussions?

The social driving forces strengthening deliberative models of political representation include problems that traditional politics face, transformations taking place in the institutions of international politics, and remedies that democratic deliberation promises to deliver. A weakening connection between politicians and citizens and declining public confidence in governments at all levels are acknowledged among the primary political problems (Andersson and Shahrokh, Chapter 4; Norris, 1999). The need for developing new means of political engagement and representation is seen to be linked to an ongoing diversification of the institutions of political representation. Contrary to 'traditional' representation theory, according to Agger et al (Chapter 3), where the only legitimate way to represent somebody is by election to one of the formal political institutions, representation in contemporary politics is seen as being performed by a range of entities: business and labour organizations, social movements and even individual public figures, who claim to provide political representation (Saward, 2006, p310).

COP15 is an apt example of a policy process where more and more of the items on the political agenda have moved beyond the scope of the sovereign nations that have traditionally defined the demos of a representative democracy. The promise of democratic deliberation through processes such as WWViews is that they can increase citizens' interest in (global) political issues, and contribute to enduring decisions which, in contrast to elected politicians' and organizations' vested interests, can reflect ideas that are not influenced by lobbyists and short-term business or electoral interests.

Ordinary citizens, in other words, can more readily be other-regarding and adopt a longer time perspective. They can thus share the burden of a political world that has become so complex that even professional politicians can remain uninformed about relevant facts and the general public's attitudes on many issues.

Methodologically WWViews is a compromise between a qualitative and a quantitative approach. In choosing the target of 100 participants per nation, DBT constructed a series of trade-offs between a 'politically legitimate' sample of the population and an affordable process that engages manageable numbers of citizens in in-depth deliberations (Agger et al). Different understandings about the size of a politically legitimate sample prevail among different observers. One benchmark is the consensus conference with its sample of 10–16 citizens, which is regarded as adequate in representing different societal viewpoints and discourses, by advocates and scholars of pTA (see e.g. Joss and Durant, 1995). An opposite benchmark is traditional surveys, in which a sample of approximately 1000 citizens is often used to provide representation over national populations, as for example in Eurobarometer surveys.

Public opinion polls with their more extensive samples are seen by many authors in this book as a main 'competitor' of WWViews in this front-line citizen consultation. In the US, for example, the organizers of the national WWViews event were challenged with a simultaneous 'Pew poll' that gave different (and less green) results about the status of public opinion on climate policy. As Schneider and Delborne (Chapter 14) argue, the Pew poll exacerbated an anxiety over WWViews' ability to be representative, because public opinion polls are generally accepted to reliably take the pulse of the country on key political issues. In order to get journalists interested in the results of WWViews, the US organizers decided to frame WWViews as an 'informed poll' (see Sclove, 2009) – a strategy that generated increased media coverage where journalistic interest was otherwise tepid.

A poll, however, is methodologically a poor proxy of WWViews and its idea of citizen deliberation. First, through the process of deliberating, the participants' ideas and attitudes are not merely expressed, but also shaped through the force of argumentation. Second, in terms of representation, the deliberative method that combines balanced and relevant information, reasoned discussion and a demographic mix of the participants who bring diverse experiences and values to the table is intended to yield results that would be similar to those generated by an exercise with another panel of demographically mixed citizens. In this way, the results of the exercise can be understood as representative, not in a simple statistical sense, but in terms of providing 'deliberative representation' or representation of different societal arguments and discourses (Agger et al; cf Renn, 2008, pp309–311). Because of this, the voices expressed through deliberation are unique in the policy process. These voices are different from polls for the reasons just reviewed; and they are different from the voices of organized interests because the latter are self-selecting, usually on the basis of pecuniary interest, whereas deliberative participants are selected by project organizers, with the rationale that their only clear interest is that of the common good (see note 3). Finally, the foregoing features establish a moral foundation for deliberation that is also unique in the policy process: neither experts nor interested parties can judge for the ordinary citizen the profound ethical matters that are bound up with issues like climate change policy, unless one wants to argue that people should not have the moral claim to make such judgements for themselves.

A number of deliberative methods comparable to WWViews, that balance qualitative and quantitative approaches, have been developed in recent years (Andersson and Shahrokh). One example is the '21st Century Town Meeting™, a methodology that has been applied in the US to involve more than 1000 residents, stakeholders and volunteers to come together to discuss common policy issues.[14] This methodology combines small-group discussion (10–12 people) with large-scale collective decision-making via the use of networked laptops, and this way aims to overcome the trade-off between the quality and quantity of discussion. Another example of mixed methodological approaches is Deliberative Polling®, a method developed by James Fishkin in the early 1990s (Abelson et al, 2003; Fishkin, 2009). The Deliberative Poll® combines a large random sample of several hundred participants with smaller-scale deliberations over a two- to three-day period. Referring to the successful experiences from these methods, Andersson and Shahrokh call for methods that move beyond the quality–quantity divide. This seems to be a promising avenue for developing the global deliberation concept further, presuming that the expenses of such processes can be contained, which is especially critical in low-income countries (see Chapter 15).

Philosophical foundations and competing models of deliberation

The methods and models of deliberation and who should participate in them are questions closely linked to an understanding of the rationales of citizen deliberation, i.e. the political and philosophical arguments that are given to support discussions within 'mini-publics' as a substitute for deliberation among the whole public. The question opens a complex set of subquestions, on which different positions can be taken, implying different models of participation.

The fundamental question is 'why deliberation' instead of some other methods of clarifying public interests and preferences. Agger et al link this question to an understanding of democracy. The deliberative democracy perspective rejects the liberal interpretations of democracy, in which citizens are regarded as having fixed interests that can be aggregated into collective decisions through methods such as voting and representation (Dryzek, 2000; Saward, 2006). Instead of proposing to 'pick up' fixed preferences, therefore, the deliberative model of democracy proposes that the best way to clarify public interests is through a deliberation process, in which the participants are amenable to changing their minds when they are presented with other viewpoints in a non-coercive way.

Second, the question of substituting general public debate and opinion with 'mini-public' debates should be distinguished from the question of substituting representative structures of government with mini-public deliberations (or 'direct democracy'). None of the contributors to this volume argue that national policy-makers or intergovernmental climate negotiations should be replaced with a public consultation. Instead, the backbone idea of WWViews to provide additional support and feedback to the official UN negotiation process (in the form of an increased sensitivity to public concerns about climate change) is echoed throughout this book.

Finally, the question of who should participate in WWViews-type citizen deliberations, and in which ways, involves conflicting ideas about the appropriate form of participation. A tension between deliberation and voting procedures is discussed by Goldschmidt et al (Chapter 5): a deliberation process aims at achieving understanding and developing an informed, common result, based on argumentation, whereas voting aims at collecting the informed preferences of the participants. For deliberation, therefore, it is enough to ensure that each type of argumentation is represented only once. For developing a valid snapshot of preferences, the voting requires a 'sampling' in a statistical sense. This implies a representation of positions in accordance with the overall distribution of positions in society, rather than a complete representation of arguments irrespective of their distribution in society. Referring to the practical difficulty of accomplishing statistically representative samples of national populations, Goldschmidt et al recommend that in future WWViews, the deliberative understanding of representation should be more prominent in the design of global deliberations. At the same time, however, the authors leave it open as to how relevant citizens' positions and arguments related to climate policy could be scoped in a more economic manner.

Following Renn's (2008, pp284–331) classification of the alternative concepts of stakeholder and public involvement,[15] the different aims related to deliberation versus voting indicate a tension between deliberative and neoliberal rationales for participation. The deliberative rationale includes consensus-seeking through argumentation, while the neoliberal rationale includes finding solutions that optimally correspond with prevailing public preferences. While reflecting on the characteristics of the WWViews model, Rask and Laihonen (Chapter 12) argue that WWViews could also be seen as an example of the anthropological concept of participation, where the main tenet is to integrate 'common sense' with the scientifically informed debate on climate policy. Parallel to the jury system in courts, even a small number of lay representatives could then be seen as representing a non-expert view of the problems under discussion.

The emancipatory concept of participation, where empowerment of historically marginalized groups is seen as a necessary condition of democracy, was not a part of the original concept of WWViews (this is criticized by Agger et al (Chapter 3), who claim that DBT was not able to find an answer to the question about the under-representation of subordinate groups of citizens). Despite this, some national WWViews partners adapted the basic model to incorporate the emancipatory concept. The Canadian project managers (Blue et al, Chapter 8), building on the work of Young (2000), argue that the criterion of representativeness can only be understood as the democratic norm of political equality if it goes beyond building a demographically representative group of participants towards also taking into account structural differences among groups in the society in question. Doing this in Canada involved several strategies. First, given the history of severe oppression of the Aboriginal population in Canada, and the higher vulnerability of this population to climate change, the organizers over-represented them in WWViews, in order to ensure that their normally suppressed or excluded voices were included in the deliberation.[16] Other strategies focused on the deliberative process rather than who was at the table. These included taking care to expressly acknowledge the diverse cultures present, and giving Aboriginals the option to meet at their own table or sit with others among the participants.

In considering different models of participation and representation, Blue et al (Chapter 8) embrace the position that all processes of representation are inherently political. Thus, the seemingly neutral step of using census data and established socio-demographic categories to recruit participants[17] can actually reproduce existing power. While Blue et al speak for the emancipatory model of participation, their argument about the inherently political nature of representation comes close to a postmodern view of representation, and a post-structuralist view of democracy that sees politics as intrinsically conflictual. Seeking consensus, in this view, will always express the hegemony of one contingent perspective (Agger et al, Chapter 3; cf Mouffe, 2000). From the postmodern perspective on participation, therefore, an open internet-based consultation, for example, might be viewed as a legitimate means to acknowledge different political perspectives on climate change, instead of forcing participants towards an (illusory) position of consensus.

The functionalist concept of participation, in Renn's (2008) typology, oriented at improving the effectiveness of political decision-making is the antipode of the postmodern concept of participation. Most contributors to this volume are more familiar with finding ways of increasing the effectiveness of citizen deliberations than with finding ways to organize larger-scale forums per se, or with creating room for unlimited deliberations and scoping of various citizen viewpoints; in this respect they resonate more with functionalist than postmodern ideas of participation.

The foregoing discussion about the different rationales and concepts of participation is related to the issue of the limits of citizen deliberation as a substitute for public opinion at large. That some models of participation contain contradictory objectives (e.g. deliberation processes aimed at producing statements expressing some type of consensus, versus voting procedures aimed at revealing prevailing preferences) indicates that practical objectives rest on distinctive theoretical and conceptual foundations.

Practical challenges

In addition to the conceptual boundaries, several authors in this volume report on the practical limitations of citizen deliberation. The difficulty of composing statistically significant citizen panels is among such limitations. Attracting enough citizens to devote their time to the full-day deliberation event also proved to be a major challenge for several national partners. Despite extensive efforts to recruit citizens, many of the national organizers did not achieve the target of 100 participants. The difficulties were most pronounced in India, the US and some European countries (for full participant numbers, see Table 1.1). The challenges at these and other sites speak to the need to devise participant recruitment strategies that will make the WWViews concept more attractive to citizens in diverse locales, without compromising the common elements in recruitment globally that make the results comparable (see Chapter 15).

Independent of whether the targeted sample size was reached, the structure of the citizen panel was another concern for national managers. Many of the authors report on their efforts to recruit specific groups of citizens (e.g. less-educated citizens, immigrants and Aboriginal people) to ensure that important societal groups are represented in the national panels, while simultaneously questioning the idea of statistical representation. This suggests that advocates of deliberative democracy can more readily agree on the necessary structures of participation (i.e. what population groups should be represented) than the adequate or optimal levels of participation within those structures (i.e. how many representatives of various population groups are necessary).

Different concerns were related to over- and under-representation of specific population groups in the WWViews panels. A tendency towards over-representation of participants with a green ideology was a general expectation and worry of the organizers. This was also expected by DBT, who therefore provided guidelines to the national managers, including the recommendation that citizens with backgrounds in environmental organizations and climate-related professions should be rejected. In practice, however, this instruction proved to be difficult to follow. In Germany, for example, the low return rate of invitations to join the WWViews event[18] left no leeway for selecting people according to their backgrounds. According to the German organizers' pre-survey to study the socio-demographic variables and attitudes of the panellists, about 10 per cent of participants were members of green organizations. To prevent a 'green overrun', the German organizers took care to distribute green and non-green participants equally between the tables, and instruct group facilitators to ensure that the viewpoints of all participants were heard. The risk of a 'green overrun' was thus managed (Goldschmidt et al, Chapter 5), but at the same time similar biases were introduced by over-representation of elderly people and participants with higher education.

The German and Finnish experiences both confirm the under-representation of less-educated citizens. Agger et al (Chapter 3) note that the problem is even more serious with citizens in marginalized social categories, e.g. the homeless, the poor and oppressed ethnic minorities. They also observe, however, that this issue affects other methodologies such as surveys, not to mention institutions of representative democracy such as national parliaments.

In some WWViews countries there were no effective means to control biases in the composition of citizen panels. An example is the Indian site of Bangalore, where the

WWViews panel was created by a non-governmental organization (NGO) recruiting local residents by word of mouth, which resulted in a group not at all representative of the diversity of the local area population (Bal, Chapter 9). Since a high standard of participant selection is central to the fairness and political legitimacy of deliberative processes (Goldschmidt et al, Chapter 5; see also Webler, 1995; Renn, 2008), more effective quality control of the recruitment process in all countries should be a high priority in future deliberations.

An interesting perspective on the practical challenges arises from the issue of timing. Understood as activity which brings together multiple elements in a given temporal space, timing emerges in several chapters of this book, and provides an access point for evaluating the dynamics and boundary conditions of global deliberative processes. The main issue discussed is whether the timing of WWViews on Saturday 26 September 2009 was optimal, both for successful deliberations and for impacting the COP15 negotiations that took place on 7–18 December 2009. Different 'windows of opportunity' for the optimal timing of WWViews can be opened, depending on the different perspectives from which the timing is viewed:

*   From the point of the view of citizen recruitment, the timing of WWViews is mostly a matter of a weekly cycle: is it possible for citizens to devote one full day of the weekend to public issues instead of family issues?
*   From the different national viewpoints, the timing is largely a matter of the festive cycle of the country: does the WWViews event coincide with national holidays, elections or other special days that restrict normal business and shift public attention towards other issues?
*   From point of view of the media cycle, the question concerns what makes an event newsworthy, attractive to journalists and competitive among other events and issues.
*   From the point of the view of policy life cycle, the question is mostly about when and how to engage policy-makers.

Saturday was deemed an unfortunate day for WWViews by the authors of two German-speaking countries and the US contributors. Goldschmidt et al (Chapter 5) consider the event's inauspicious timing to be one of the main reasons for the low (less than 1 per cent) return rate of the citizens invited to the German WWViews. Both the Austrian (Bechtold et al, Chapter 6) and US authors (Schneider and Delborne, Chapter 14) also deemed Saturday a bad choice in terms of media attention, since it typically is a slow day for most news organizations.

In terms of national holidays, the complexity of the situation increases with the number of countries involved. The DBT staff consulted a global festival calendar before setting 26 September as the day of the global consultation. The level of detail of such calendars, however, may not be adequate to take into account less prominent events. Surprises with overlapping national events were reported from Uruguay, where

WWViews day coincided with the celebration of National Heritage Day (Bortagaray et al, Chapter 10); India, where it fell on a major Hindu festival of Navrartri (Bal, Chapter 9); and Germany, where the WWViews event took place just one day before the German national elections.

A more fundamental timing challenge is the optimal duration between the deliberation and the critical decision day(s) for the issue being considered. The pros and cons of early and late activity are philosophically expressed in Collingridge's (1980, p19) 'dilemma of the social control of technology':

> ... *attempting to control a technology is difficult, and not rarely impossible, because during its early stages, when it can be controlled, not enough can be known about its harmful consequences to warrant controlling its development; but by the time these consequences are apparent, control has become costly and slow.*

Applying this wisdom to the timing of WWViews means that impacting the UN climate change negotiation process would be most influential in its early stages (maybe one or two years before the top meeting), when the issues are truly open, but the risk is that there is no guarantee of the relevance of those issues a year or two later. On the other hand, if a later stage is preferred (as it was with WWViews), to wait for the issues of the negotiation to be settled, then the risk is that affecting negotiations gets difficult, if not impossible.

While the dilemma of timing seems a classic question of trading off and optimizing two opposing choices (characterized by Collingridge's dilemma), an interesting perspective on this question emerges from the disparate rhythms of media and policy cycles. Thinking about the months or even years preceding the preparation of the COP15 negotiation, it is evident that policy cycles related to international decision-making are extremely long. Thinking about how impulsively media selects its news stories, on the other hand, and how quickly those stories pass through the media spotlight, a radically shorter time span is introduced. Given its relatively late timing, the WWViews project was obliged to rely on media breakthroughs more than building institutional cooperation with national and international policy-makers (see Schneider and Delborne, Chapter 14).

The role of media

Reflecting the importance it placed on media coverage, DBT prepared both global and national media strategies. DBT was appropriately pleased with TV coverage in Denmark in a widely viewed programme slot the day after WWViews, but several partners' expectation that WWViews would be a 'seller' due to its topicality and global scope[19] were not met by large-scale and substantive coverage, neither in partner countries nor at COP15.

Media coverage nonetheless varied in both quality and extent. In Australia, for example, the organizers were successful in obtaining prime-time news coverage on the day of the event on a national television channel, and articles in several leading state-based newspapers; they also achieved substantial regional news coverage by focusing on the personal stories of WWViews participants (Herriman et al, Chapter 13). In Denmark, the press coverage of WWViews was not very extensive, but the project was reported in a five-minute feature on the prime-time television news on Radio Denmark the day after the citizen consultation (Agger et al, Chapter 3), and some two dozen articles in online and print newspapers were published in the three weeks following the event.[20] The Danish organizers were also contacted by international media, including the Japanese public broadcasting company NHK, which produced a documentary film of the Danish, Indian, Maldivian, US and Japanese experiences of WWViews. A new wave of media attention followed the COP15 meeting in Copenhagen on 7–18 December 2009. In India media coverage of WWViews was practically nil despite efforts to use a university-based public relations office to help in dissemination (Bal, Chapter 9). In the US sites (Arizona, California, Colorado, Georgia, Massachusetts) the organizers were unable to garner significant media attention, even though media coverage expanded considerably after a professional media consultant was hired to help disseminate the results and search for news holes (Schneider and Delborne, Chapter 14).

Three main issues stand out in the discussions of media in this volume. First, the primary explanation for the low level of WWViews media coverage is lack of resources, primarily meaning funding, but also including professional competence. Another issue is how to frame the story of WWViews. Is it about climate change? Democracy? Public opinion? Third, and more philosophically, there is the question of the purpose of media coverage.

Despite significant efforts, DBT was unsuccessful in attracting an institutional partner with significant access to global media (e.g. major foundations, official United Nations bodies or major international media such as the BBC), no doubt in part because resources were pursued during the worst global economic downturn since the depression of the 1930s. This was perhaps the greatest limitation in the WWViews media initiative (Agger et al, Chapter 3), because it put more responsibility on national sites to generate coverage for a story that probably has more appeal as a global phenomenon.

Addressing media strategy at a national level, many of the authors in this book acknowledge that effective dissemination is resource-intensive and requires professional communication competences. Herriman et al (Chapter 13), for example, on the basis of their successful experiences with the Australian WWViews, suggest that dissemination should be considered as a 'whole-of-project endeavour' and not just something that happens once the results are out. They also emphasize that dissemination should start early, adopt a flexible approach that provides tailored information to specific target groups, and allocate resources to post-event follow-up. While the Australian partner was successful in doing this, most partners simply lacked the resources needed for such an undertaking. Still, in many countries considerable resources were invested in media outreach without much success, raising a question about the appropriate balance between effort and pay-off. While no simple quantified answer can be given to this question, many of the authors in this volume concur that professional media skills are critical, and often lacking among the academics and deliberative practitioners who constituted the leadership of most national teams.

Framing WWViews is another challenge. Should WWViews be framed as a deliberative democracy story or a climate change story? In other words, is the core issue of WWViews about a new participatory method or about its results, reporting on citizens' opinions about climate policy? Should the continuity or discontinuity of WWViews with other processes and results be emphasized? Are there any particular conflictual interests or viewpoints that should be included in communications about WWViews? These are some of the questions that need to be answered while communicating about WWViews or other deliberative projects.

According to Schneider and Delborne, WWViews was reported mainly as a 'climate change story', at least in the US mass media. Since there is great competition for attention towards climate change stories, and more generally little or cyclical attention to environmental stories (as compared to other stories such as economic crises or health issues, see e.g. Anderson, 1997; Cox, 2006), the window of opportunity to WWViews for gathering media attention was small. In Uruguay (Bortagaray et al), the project team recognized that climate change received little attention in the country's media, and consequently focused their outreach on the method of citizen consultation as a novel undertaking in the Uruguayan context. Sensible though this strategy was, it nonetheless garnered little attention.

Media scholars have identified several criteria of newsworthiness (e.g. Anderson, 1997; Boykoff and Boykoff, 2004; DeLuca, 2006). According to Schneider and Delborne, WWViews successfully met two of them: novelty (citizen deliberations had occurred before but not on the scale of WWViews) and continuity (climate change is an ongoing story); but failed to meet many other important criteria, such as clarity (or lack of ambiguity), narrative and event-centred focus (is the story about WWViews day or COP15?), meaningfulness, emotionality and conflict.

The absence of dramatic conflict probably posed the greatest constraint in selling the WWViews story to the media. DBT's role as a neutral and non-partisan organization partially explains why the more aggressive media strategies typically deployed by advocacy organizations were not adopted in communications about WWViews (Agger et al; Schneider and Delborne). Quite interestingly, however, in the US a conflict narrative was devised by the media consultant who connected WWViews to a concurrent Pew Center poll on climate change that had grabbed headlines with the datum that Americans' interest in or concern about climate change had fallen nearly 20 per cent in two years (Pew Research Center, 2009). Sclove's 2009 article entitled 'Why the polls on climate change are wrong' directly challenged the results of the Pew poll – by referring to the contrary results from WWViews – and in that way yielded significant attention, at least, in green and liberally oriented media (Schneider and Delborne).

An orientation to media coverage, and especially strategic usage of conflict framing, can prove a double-edged sword that risks doing harm to the original deliberative concept of WWViews. As Schneider and Delborne suggest, by presenting WWViews as an 'informed poll' that stands in contrast to the Pew Centre's poll and indicates (Americans') uniform support for strong climate policy, the organizers of WWViews risk both narrowing the broad range of issues deliberated in the WWViews events and committing themselves to one particular partisan position. Since one aspiration of WWViews partners was to contest the stale 'pro' and 'con' format in current climate policy conflict by presenting the results of deliberations by diverse groups of everyday citizens, the conflict-oriented media strategy can be risky.

To conclude, while reflecting about the role and purpose of media coverage in WWViews, the authors in this book make two very different kinds of suggestions. First, media as a 'friend' could more effectively be mobilized by expanding media strategies from the current focus on who to contact, to instead concentrating on how to get coverage by strategic framing. Second, media's role as a potential 'enemy' of deliberation calls for shifting attention from impacting through mass media towards impacting through more direct policy pathways, such as building partnerships with policy-makers or institutions and emphasizing public outreach and education instead of public attention per se (Schneider and Delborne; see also Agger et al; Rask and Laihonen).

Tracking of policy impacts

As found in studies on consensus conferences (Agger et al; see also Einsiedel et al, 2001), a narrow conception of political outcomes will probably overlook most political impacts of deliberative processes. Several articles in this volume that track the immediate policy impacts of WWViews sustain this view.

The direct political influence of WWViews was disappointing to many, who understandably hoped that the project goal 'opportunity to influence global climate policy' would translate into actual influence. Nine out of ten WWViews participants urged their national delegations to reach a new, binding climate change agreement at COP15.

Despite the limited direct impacts at COP15, there are already signs of broader policy effects. Among these indirect political influences are procedural, educational and politico-symbolic impacts. Procedural impacts refer to the way that different stakeholders view deliberative practices, and how the latter become institutionalized in different decision-making contexts. In Uruguay, for example, the organization of the WWViews event was integrated with the university curriculum of first-year biology students to provide them with the 'tools required to incorporate reflection about the societal role and concept of science as part of their professional practice' (Bortagaray et al). Another example is from Saint Lucia, where methodological aspects of WWViews were soon afterwards applied in an Energy Symposium organized by a government office that collaborated with the local WWViews partner. Similar methods have been considered for a Youth Environment Forum (Charles et al, Chapter 11). Clearly a longer time perspective is needed to fully understand the wider implications of WWViews both nationally and internationally. What the evidence to date strongly suggests, however, is that the

WWViews process can generate increasing attention to, and interest in, deliberative processes internationally.

The educational impacts include what participants and the public learned. In a participant exit survey conducted in 16 countries and 21 WWViews sites, a majority of the respondents reported that WWViews 'significantly influenced their opinion about climate change'; they also became motivated to follow climate political debates and search for more information about climate change.[21] Like policy impacts, more indirect learning effects of WWViews can best be understood over longer time horizons.

The politico-symbolic impact refers to the symbolic value of the WWViews event as a demonstration of the ability of ordinary citizens to take part in complex decision-making (Agger et al). A telling finding in this respect is that participating citizens themselves see WWViews as a meaningful contribution to political decision-making.[22] The very high levels of satisfaction are a strong argument for continued development of global democratic deliberations along the lines of the structured face-to-face deliberation model of WWViews.

In thinking about what explains diverse impacts and local adaptations of WWViews in different countries, many factors can be considered, ranging from structural and cultural issues to more dynamic matters of actor dynamics in the policy arenas, and organizational competencies to conduct and absorb deliberative processes. In their account of such factors, Rask and Laihonen (Chapter 12) propose a framework building on the tradition of 'policy transfer' studies, especially Sabatier's (1988) advocacy coalition framework (ACF) and related theoretical work (Fenger and Klok, 2001; Stone, 2004; Heiskanen et al, 2009). This framework focuses attention on the receiving institutional context; the carriers of the models; and the role of 'translation as an active process' (the last referring to the idea that policy transfer is an active process involving learning, strategic and tactical play by various players).

The impact of institutional context on the local implementation of WWViews is discussed by several contributors to this volume. Among the countries involved in WWViews (see Table 1.1), there is a large variation both in terms of their socio-economic backgrounds and their institutional frameworks and policy systems. The main socio-economic distinction (as in COP15 negotiations) is between industrial and developing countries. In countries such as India, poverty, illiteracy, gender gap, caste distinctions and high diversity between regions are factors that can constrain efforts to organize deliberation processes based on the ideals of equal participation by all relevant societal groups (Bal). While such constraints can be less pronounced in Western industrial countries, other factors operate there, such as different religions, languages, minority and ethnic groups, nationalities, and educational and professional backgrounds, which complicate the management of deliberative processes. The latter group of factors, however, are less significant in delimiting citizen contributions to policy deliberations and more challenging for organizers, who must cover all relevant groups and ensure social representation. Less clear is whether any of the socio-economic factors directly affect the way that deliberation processes are adopted in their local policy contexts.

More important, perhaps, in explaining the impact of policy context on the transfer and adaptation of deliberative processes, are differences between institutional frameworks and systems of public accountability that exist with regard to climate policy and science. Bal cites the concept of 'civic epistemology' from Jasanoff (2005, p255) to detect such impacts. Civic epistemology, according to Jasanoff (2005) '... refers to the institutionalized practices by which members of a given society test and deploy knowledge claims used as a basis for making collective choices'. In considering WWViews, different 'civic epistemologies' prevail in different countries, and therefore impact the way that the deliberative process and its results come to be seen as authoritative information and a legitimate part of political decision-making. Components of civic epistemology include factors such as local policy and assessment styles, availability of science, policy analysis methods, local issues and knowledge, and media and public perception of policy issues (Bal; see also Jasanoff, 2005; Miller, 2005).

In terms of their 'civic epistemologies', countries can roughly be classified into supportive, transitional and resistant towards deliberative democracy. For example, public participation and deliberation is institutionally embedded in Denmark. Operating in a context accustomed to deliberative processes means lesser investment in justificatory measures, and conducting and communicating them in a business-as-usual manner. Even a supportive policy context, however, does not guarantee high policy impact, which is observed by Agger et al from the experience of rather limited direct policy impacts of consensus conferences in Denmark.

Many of the WWViews partners can be characterized as transitional countries in terms of their tradition of citizen deliberation processes. In Uruguay, for example, WWViews was the first-ever large-scale exercise in deliberative democracy. The Uruguayan context includes a civilian democratic tradition with a mandatory vote and a participatory attitude towards conventional political issues, but the government has only recently started to pay attention to science, technology and environmental policy issues. The advantage of this situation in organizing WWViews-type deliberations, according to Bortagaray et al, is that it helps to establish institutional contacts '... in a friendly political environment towards S&T'. As their case study indicates, a high number of contacts were made with high-level political and scientific actors. Despite this political support, however, attracting media attention to climate change-focused WWViews proved to be challenging. In sum, however, the Uruguayan WWViews case provides a very positive example of how a successful experience with a new type of deliberative process both can help transfer such processes into new policy contexts (a consensus conference on nuclear energy was organized in 2010) and find creative ways to combine them with other existing activities (the WWViews process was creatively combined with a university seminar on biology and society).

We close by comparing Finland and India as two different examples of policy contexts where the traditional system and culture of policy-making causes resistance to deliberative democratic initiatives. In terms of its policy culture and 'civic epistemology', Finland can be characterized as an 'exclusive-corporatist' state (Pelkonen,

2008). One important factor is that the Finnish understanding of civic participation means participation in civic organizations (Savikko and Kauppila, 2009). Even if final decision-making takes place in the closed circles of specific governmental and parliamentary committees, Finnish regulatory processes are typically accompanied by extensive stakeholder consultations. This means that issues of policy-making are generally deliberative and participatory, but not involving citizens or consumers outside organized interest groups or organizations (Rask and Laihonen).

In India, the opportunities for citizen deliberation seem even more restricted. According to Bal, science policy in India is trapped in the 'deficit model' that shifts responsibility for structural deficiencies to individual citizens, and the primary objective of official science policy is '… developing a scientific disposition in the people and educating them "to the marvels that science can bring"'. Considering the different constraints that the resistant policy context causes for the local adaptation of deliberative processes, different remedies are proposed by the Finnish and Indian authors. While in Finland an alliance between WWViews and local NGOs is suggested as a measure to increase the visibility and policy impact of WWViews, the Indian remedy envisions a more distant prospect of a radical change in thinking about the governance of science and technology that will create the space for citizen deliberations in the realm of governance practices.

Scholars interested in democratic deliberation and technology assessment will find in this volume rich empirical documentation of an unprecedented event, analysed in terms of the literature in these fields built up since the late 1980s. Practitioners will see how specific approaches fared in a diverse array of contexts. The sustained interactions of people situated in both professional arenas, and employed in the public, private and non-profit sectors, have informed the questions and analyses presented here since early in the WWViews project. In terms of audience, however, our foremost hope is that this volume will be read by students. The editors and authors have considerable experience with students who have learned about deliberative practices in the classroom and tested them in their later professional and civic lives. We therefore see students as archetypical readers of this book: people who are expanding their horizons, developing analytical and people skills, in some cases preparing for their careers, and in all cases becoming citizens in a world that is likely to build on the many initiatives over the past quarter-century to develop new modes and meanings of democracy.

Notes

1   The UNFCCC was formally created in 1994 as a product of the United Nations Conference on Environment and Development in Rio de Janeiro (Brazil) in 1992. The Copenhagen summit was the 15th meeting of the Conference of Parties to the UNFCCC, hence the acronym COP15.

2   This phrase varies only slightly from the title of John Dryzek's *Deliberative Global Politics* (2006), to which the editors owe a considerable intellectual debt.

3   We use the terms 'ordinary people' and 'citizens' in this volume to refer to individuals who have no significant expertise or pecuniary interest in climate change. An atmospheric chemist, for example, might be an ordinary person in many senses of the word, but her expertise would disqualify her from that category for purposes of being a citizen participant in WWViews. A less obvious example would be a passionate volunteer for a climate change advocacy organization. While conventional notions of expertise tend to overlook this category, there is indeed a considerable and growing number of citizen experts on most policy issues (Worthington, 2007; Rask, 2009), and they were an obvious presence among the tens of thousands of people who went to Copenhagen to observe and influence COP15. While participants in deliberative exercises like WWViews apply to the project managers to be considered for inclusion in the event, and in that sense are self-selecting, the selection criteria exclude applicants judged to be experts or interested parties. This is the key to their categorization as 'ordinary people' and the core difference between them and just about everyone else who shows up at events like COP15.

4   The results of the Snekkersten workshop are reported in a Climate Arena blog, see http://blogit.kuluttajatutkimus.fi/ilmastoareena.

5   This section is largely based on the WWViews Policy Report (Bedsted and Klüver, 2009) in which the results of the project are summarized. The report is freely downloadable at www.wwviews.org.

6   All the information material for WWViews was produced in English and translated into local languages for non-English speakers.

7   The information booklet and its translations into ten different languages is available at www.wwviews.org.

8   There are, however, some major differences between countries. As the world's largest emitter of greenhouse gases and the largest fast-growing economy, China is of particular interest. Of the Chinese participants, 45 per cent supported actual emission reductions in their own country, whereas a small majority (another 52 per cent) supported limitation of the growth of emissions as the target for growth economies. In contrast, the majority (72 per cent) of participants from other fast-growing economies supported actual emission reductions. (See Bedsted and Klüver, 2009, p19.)

9   See the previous note.

10 Author names without indication of the year of publication refer to chapters in this volume.

11 pTA is a class of methods that actively involves 'social actors' in the assessment of socio-technological issues (see e.g. Decker and Ladikas, 2004; Joss and Bellucci, 2002).

12 The following chapters examine 11 of 38 WWViews countries, leaving a large majority outside the book's scope. This unfortunately includes countries lacking a participatory tradition, such as China, Russia and several African countries.

13 The statistical and mathematical treatment of representation in this context is actually lacking, and deserves more systematic attention in future. For example, the mere size and diversity of the populations in some countries, such as China and India, make it challenging to find any kind of statistically representative way of representing the populations for deliberative purposes.

14 For further information, see http://americaspeaks.org.

15 Renn (2008, p303) distinguishes between six concepts of stakeholder and public involvement: functionalist, neoliberal, deliberative, anthropological, emancipatory, postmodern.

16 Aboriginals constitute 3.8 per cent of Canada's population, but were 10 per cent of the WWViews participants. These slots were taken from the most populous provinces of Quebec and Ontario.

17 In the case of WWViews, the categories used to assemble a balanced group of participants were age, sex, educational and professional background, and living area. Religion, ethnicity, language and immigrant status are examples of other categories that were applied in some countries.

18 In Germany, a random selection method was used to recruit citizens. Of the 9000 invitation letters sent to citizens, 135 responded positively (1.5 per cent); 81 actually attended the event (0.9 per cent); see Goldschmidt et al.

19 In Finland, for example, a research trainee was recruited a month before the 26 September 2009 event with the task of conducting media follow-up; however, it soon became evident that the sparse media coverage provided little to pursue, so her duties were changed.

20 WWViews media report posted by Mads Petersen from DBT on 6 November 2009.

21 The exit survey consisted of positive statements to which answers were selected from a seven-point scale ranging from fully agree (1) to fully disagree (7), number 4 indicating the position of neither agreeing or disagreeing. The average response rate for the claim 'the WWViews process significantly influenced my opinion about climate change' was 2.75; 1.84 for the claim '... motivated me to follow the political debates on climate change in the future' and 1.92 for the claim '... motivated me to search for more information on climate change issues in the future'.

22 In the seven-point scale (see note 21) the average figure for the claim 'The results achieved are a meaningful contribution to political decision-making on climate change' was 1.99, and correspondingly 1.34 for the claim 'It's beneficial to continue dialogue processes such as the WWViews project in the future'.

References

Abelson, J., Forest, P.-G., Eyles, J., Smith, P., Martin, E. and Gauvin, F.-P. et al (2003) 'Deliberations about deliberative methods: Issues in the design and evaluation of public participation processes', *Social Science and Medicine*, vol 57, no 2, pp239–251

Anderson, A. (1997) *Media, Culture and the Environment*, UCL Press, London

Bedsted, B. and Klüver, L. (eds) (2009) *World Wide Views on Global Warming – From the World's Citizens to the Climate Policy Makers,* Policy Report, The Danish Board of Technology, Denmark, www.wwviews.org/files/AUDIO/WWViews%20Policy%20Report%20 FINAL%20-%20Web%20version.pdf, accessed 18 April 2011

Boykoff, J. and Boykoff, M. (2004) 'Journalistic balance as global warming bias: Creating controversy where science finds consensus', *Extra! The Magazine of Fairness and Accuracy in Reporting*, www.fair.org/index.php?page=1978, accessed 29 December 2010

Collingridge, D. (1980) *The Social Control of Technology*, Frances Pinter, London

Cox, R. (2006) *Environmental Communication and the Public Sphere*, Sage, Thousand Oaks, CA

Decker, M. and Ladikas, M. (eds) (2004) *Bridges between Science, Society and Policy: Technology Assessment – Methods and Impacts*, Springer Verlag, Berlin

DeLuca, K. M. (2006) *Image Politics: The New Rhetoric of Environmental Activism*, The Guilford Press, New York

Dryzek, J. S. (2000) *Deliberative Democracy and Beyond: Liberals, Critics, Contestations*, Oxford University Press, Oxford

Dryzek, J. S. (2006) *Deliberative Global Politics: Discourse and Democracy in a Divided World*, Polity Press, Cambridge

Einsiedel, E. F., Jelsøe, E. and Breck, T. (2001) 'Publics at the technology table: The consensus conference in Denmark, Canada, and Australia', *Public Understanding of Science*, vol 10, no 1, pp83–98

Fenger, M. and Klok, P.-J. (2001) 'Interdependency, beliefs, and coalition behaviour: A contribution to the advocacy coalition framework', *Policy Sciences*, vol 34, no 2, pp157–170

Fishkin, J. (2009) *When the People Speak: Deliberative Democracy and Public Consultation*, Oxford University Press, New York

Guston, D. (1999) 'Evaluating the first U.S. consensus conference: The impact of the citizens' panel on telecommunications and the future of democracy', *Science, Technology & Human Values*, vol 24, no 4, pp451–482

Heiskanen, E., Kivisaari, S., Lovio, R. and Mickwitz, P. (2009) 'Designed to travel? Transition management encounters environmental and innovation policy histories in Finland', *Policy Sciences*, vol 42, no 4, pp409–427

IPCC (2007) Climate Change 2007: Synthesis Report. Contribution of Working Groups I, II and III to the Fourth Assessment Report of the Intergovernmental Panel on Climate Change. Core Writing Team, Pachauri, R.K. and Reisinger, A. (eds) IPCC, Geneva, Switzerland, p104 http://www.ipcc.ch/publications_and_data/ar4/syr/en/contents.html

Jasanoff, S. (2005) *Designs on Nature: Science and Democracy in Europe and the United States*, Princeton University Press, Princeton, NJ

Joss, S. (1998) 'Danish consensus conferences as a model of participatory technology assessment: An impact study of consensus conferences on Danish Parliament and Danish public debate', *Science and Public Policy*, vol 25, no 1, pp2–22

Joss, S. and Bellucci, S. (eds) (2002) *Participatory Technology Assessment: European Perspectives*, Centre for Study of Democracy, London

Joss, S. and Durant, J. (eds) (1995) *Public Participation in Science – The Role of Consensus Conferences in Europe*, Science Museum with the Support of the European Commissions Directorate General XII, London

Kleinman, D. L. and Powell, M. (2007) 'A toolkit for democratizing science and technology policy: The practical mechanics of organizing a consensus conference', *Bulletin of Science, Technology & Society*, vol 27, no 2, pp154–169

Miller, C. A. (2005) 'New civic epistemologies of quantification: Making sense of indicators of local and global sustainability', *Science, Technology & Human Values*, vol 30, no 3, pp403–432

Mouffe, C. (2000) *The Democratic Paradox*, Verso, London

Norris, P. (1999) *Critical Citizens: Global Support for Democratic Government*, Oxford University Press, Oxford

Pelkonen, A. (2008) *The Finnish Competition State and Entrepreneurial Policies in the Helsinki Region*, PhD thesis, University of Helsinki, Helsinki, http://helda.helsinki.fi/handle/10138/23372

Pew Research Center (2009) *Fewer Americans See Solid Evidence of Global Warming: Modest Support for 'Cap and Trade' Policy*, The Pew Research Center, Washington, DC, http://people-press.org/report/556/global-warming, accessed 3 January 2011

Rask, M. (2009) *Expansion of Expertise in the Governance of Science and Technology*, Lambert Academic Publishing, Köln

Renn, O. (2008) *Risk Governance: Coping with Uncertainty in a Complex World*, Earthscan, London

Sabatier, P. A. (1988) 'An advocacy coalition framework of policy change and the role of policy-oriented learning therein', *Policy Sciences*, vol 21, no 2–3, pp29–168

Savikko, R. and Kauppila, J. (2009) '*Kyl se sit niissä verkostoissa tapahtuu se päätöksenteko*' – *Kansalaisvaikuttaminen ilmastopoliittiseen päätöksentekoon Suomessa* (draft 19 October 2009). Citizens' global platform: Marginal voices project, www.globalplatform.fi/files/kansalaisvaikuttaminen_ilmastopoliittiseen_paatoksentekoon%E2%80%A6.pdf, accessed 18 April 2011

Saward, M. (2006) 'The representative claim', *Contemporary Political Theory*, vol 5, no 3, pp297–318

Sclove, R. (2000) 'Town meetings on technology: Consensus conferences as democratic participation', in D. L. Kleinman (ed) *Science, Technology and Democracy*, State University of New York Press, Albany, NY

Sclove, R. (2009) 'Why the polls on climate change are wrong', *The Huffington Post Blog*, www.huffingtonpost.com/richard-sclove-phd/why-the-polls-on-climate_b_331896.html, accessed 25 April 2011

Sclove, R. (2010) 'Reinventing technology assessment: A 21st century model', Science and Technology Innovation Program, Woodrow Wilson International Center for Scholars, http://wilsoncenter.org/topics/docs/ReinventingTechnologyAssessment1.pdf, accessed 18 April 2011

Stone, D. (2004) 'Transfer agents and global networks in the "transnationalization" of policy', *Journal of European Public Policy*, vol 11, no 3, pp545–566, http://wrap.warwick.ac.uk/1742, accessed 18 April 2011

Webler, T. (1995) 'Right discourse in citizen participation: An evaluative yardstick', in O. Renn, T. Webler and P. Wiedemann (eds) *Fairness and Competence in Citizen Participation: Evaluating New Models for Environmental Discourse*, Kluwer Academic Publishers, Dordrecht

Worthington, R. (2007) 'Community-based research and technoscience activism: A report on the Living Knowledge 3 Conference', *Science as Culture*, vol 16, no 4, pp475–480

Young, I. M. (2000) *Inclusion and Democracy*, Oxford University Press, Oxford

2

The story of WWViews

Bjørn Bedsted, Søren Gram and Lars Klüver

World Wide Views on Global Warming (WWViews) was the first-ever globe-encompassing deliberative citizens' consultation. It offered nearly 4000 citizens in 38 countries the possibility to communicate their views on the COP15 climate negotiations in Copenhagen to their policy-makers.

After day-long deliberations on 26 September 2009, it became clear that there was a strong mandate from citizens to take fast and strong action at COP15. This information would not have been available to policy-makers had WWViews not taken place.

Box 2.1 Programme for 26 September 2009

All meetings followed the same schedule: the 100 citizens, divided into tables of five to eight people, were led by group moderators through a programme divided into four thematic sessions and a recommendation session.

During the thematic sessions, citizens voted on alternative answers to a total of 12 questions. Each thematic session was introduced by the facilitator and an information video. The participants then engaged in moderated discussions at their tables, and each thematic session concluded with citizens casting their votes anonymously on two to four questions. Votes were counted and immediately reported to www.wwviews.org.

During the recommendation session, citizens wrote in their own words what they believed to be the most important recommendation to pass on to COP15 negotiators. Each table produced one recommendation and all citizens then voted for the ones from all the tables that they found most important, resulting in a prioritized list of recommendations, also reported to www.wwviews.org.

Box 2.2 WWViews Alliance

WWViews is structured as a global alliance of institutions, including public councils, parliamentary technology assessment institutions, civil society organizations and universities. Over 50 national and regional partners in 38 nations are part of the WWViews Alliance. Together, they facilitated 44 deliberations on 26 September 2009.

We would like to share here some of our motivations for introducing citizen participation to global policy-making, as well as some of the considerations behind the methodological choices we made as developers and coordinators of WWViews.

Responding to new global challenges

The regulation of technologies and environmental issues is increasingly coordinated through transnational and international negotiations, reflecting a growing awareness of global interdependency. As more policy-making becomes global in scale, new methods are needed in order to include and engage citizens worldwide in the policy-making processes. The challenge, as we see it, is to close a widening democratic gap between policy-makers and citizens, thereby increasing the latter's ownership of decisions made about current global issues.

WWViews is a response to this challenge. It was initiated by the Danish Board of Technology (DBT), which is the Danish parliamentary technology assessment institute, and thus an adviser to the Danish Parliament and other political decision-makers in Denmark. Especially in north-western European countries, citizen participation has increasingly become a natural element of technology assessment (TA) during the last 25 years, offering politicians, scientists and other stakeholders a better understanding of the public's grasp of new scientific and technological developments as well as its wishes for the direction such developments should take. At DBT, we have been strong advocates for citizen participation and its inclusion in the technology assessment toolbox. We have developed and implemented a wide range of methods for involving citizens in political decision-making processes – locally, nationally and on a European scale. Our rationale for doing so is partly democratic ('it is only fair to consult those who are going to be affected by political decisions before they are made') and partly functional ('political decisions have a better chance of being sustainable and properly implemented if in tune with the public').

Since DBT is internationally renowned for its leadership in the development and implementation of citizen participation, it was natural for us to take the lead in introducing citizen participation at the global level. NGOs have already run several campaigns with a global scope. As we see it, it was a logical step and only a matter of time before citizen participation was also applied to address the new global challenges. The COP15 climate negotiations in Copenhagen were an obvious opportunity to do so. Decisions made (or not made) in Copenhagen and beyond would obviously affect all world citizens to some extent. In a debate heavily dominated by scientists, politicians, industrial lobbyists and professional NGOs, each with their own agenda and often claiming to represent the citizens, there was a need to systematically include the views of everyday citizens in the debate.

Building the WWViews Alliance

We decided in 2007 to initiate WWViews and went looking for partners. Our initial criterion of success was to have at least 15 countries participating. Within three weeks of introducing the idea to our international network of organizations (mainly in Europe, and some in Asia and North America), sharing our understanding of technology assessment and citizen participation, we had ten partners signed up and another ten hoping to do so. This early confirmation from close colleagues made us confident that we would succeed in establishing the project and we could concentrate on extending our network to new countries, especially developing countries and emerging economies. It took much time and effort to identify qualified partners in these countries, but the general response was that organizations in all parts of the world easily understood and accepted the purpose and outline of WWViews. The biggest barrier for signing up more partners was the fact that they had to come up with their own funding for participating. While we did manage to raise some funding from various sides (i.e. the Danish and Norwegian foreign ministries), most partners ended up paying for all or most of their own participation. Regretfully, some partners had to leave the project because of lack of funding. We ended up with 44 partners in 38 countries but could have covered more than 55 countries if only the funding had been available.

Having taken the initiative, we were nevertheless eager to promote common ownership of WWViews among our partners. The organizational structure, the 'WWViews Alliance', reflected this wish to delegate responsibility and ownership to project partners, while locating the 'WWViews secretariat' at DBT. We had many discussions in DBT about balancing the wish for common ownership with the wish for effective coordination. We formed various workgroups within the WWViews Alliance addressing different issues such as the overall methodology, questions to citizens, information material, media relations and contact with decision-makers. This input from project partners (exchanged via our intranet) proved to be essential, but it was also clear to us that a relatively centralized decision-making process was necessary in order to ensure the coherence and progression of WWViews. Luckily, our partners shared this understanding.

Uniformity and global diversity

We started in late 2007 to design a method that would make it feasible for potentially all countries in the world to participate. Economic limitations had a large impact on the method design since we wanted to make it economically feasible for as many countries as possible (both rich and poor) to participate. While keeping the design as cheap and simple as possible, we wanted it to build on well-established praxis for citizen participation (informed citizens, face-to-face deliberations, etc). So the design had to balance a number of considerations while delivering pragmatic solutions. For example, we wanted on the one hand to allow the participating citizens to express their views as freely as possible, and on the other hand these expressions should be focused enough to allow for clear communication to decision-makers.

The cornerstones of the WWViews method were laid down in a workshop with some of the first partners to join WWViews and developed in further detail during 2008 and 2009. As method, WWViews is a hybrid of a handful of well-tested citizen participation methods previously used at the local, national and regional levels, such as the voting conference and interview meeting (voting on choices), citizen hearing (table brainstorm and meeting priority-setting on recommendations), the consensus conference (principles of composition of the information materials), focus groups (test of information materials and questions). The method falls into the same method cluster as the citizen summit (America*Speaks*) and the Deliberative Poll® (James Fishkin), but with some important differences – for example, the procedure for selection of citizens is more elaborate than in the citizen summit, and less elaborate than in the Deliberative Poll®; also, the WWViews method is specifically designed for simultaneous use at smaller meetings, arranged by many partners, with many languages on very many sites.

One of the key issues debated in the workshop was whether or not tools mostly developed in Western societies were applicable to other parts of the world. Would citizens and policy-makers understand and accept the 'rules of engagement' in countries without similar traditions of citizen participation? Flexibility in order to accommodate cultural differences had to be balanced against the wish to make results comparable across the world. Would all citizens have to answer the same questions? Should citizens with very diverse practical experience and educational background be addressed and engaged in the same way? Based upon the experience that other participatory methods, such as the consensus conference and the scenario workshop, had 'travelled well' from Denmark to other countries such as Taiwan, Korea, Japan and Chile, we decided (despite concerns raised at the workshop) that a highly uniform approach was probably feasible in a global context, and was in fact needed in order to make the comparison of results trustworthy and easy to communicate to policy-makers. This decision was reflected in many of the choices we made later about project design.

Questions asked

For example, we decided that the questions posed to the citizens along with the answer options should be predefined and identical in all countries. This decision runs counter to one of the ideals in citizen participation, namely that citizens should be allowed to define their own agenda in order not to be framed too hard. In this case, however, the agenda was already set by the climate negotiations, and we felt that it would be impossible to communicate in any politically relevant way a huge multitude of agendas defined by citizens themselves. To some extent, this problem was a matter of scale. In a national context, designers of citizen participation also face the dilemma of choosing between the wish to empower citizens in a bottom-up approach and the wish to deliver precise input targeted at a policy-making process. At a national scale this dilemma can be solved by combining survey-type results with broader qualitative expressions. When moving to a global scale, it arguably becomes more difficult to capture the diversity of the qualitative input (the amount of data is immense) and translate it into clear messages of global political relevance, which is what we were aiming for with WWViews. If we neglected to do so, we would fail in our responsibility to give participating citizens the best possible chance of getting their views across to policy-makers. We therefore decided to frame a set of survey questions quite strongly and as a counterweight to this framing, through a 'recommendation session', allow citizens to put in their own words their top recommendations to the climate negotiators. In order to be able to handle the vast amount of expressions from the recommendation session, these recommendations were prioritized on site by the citizens in each country. Results from this session were used extensively in the final policy report to qualify the analysis of the voting on the predefined questions in the first four sessions of the programme. They also resulted in policy recommendations on issues not addressed in the predefined questions.

It is often debated among citizen participation practitioners whether or not one should strive to bring deliberations about the issues at hand as close as possible to the daily lives of the citizens. While open to the wish of some partners to also focus on individual choices, we were faced with the problem of scale and with the truly global nature of the UN negotiations. For example, it was felt that a test of citizens' personal will to give up concrete benefits in order to curb climate change would deliver politically relevant results. It proved meaningless, however, to phrase such a question in a concrete manner both to an American businessman and an Ethiopian farmer ('would you be willing to give up driving your car to work in order to curb climate change?'). Alternatively, one could phrase the question in a more general way ('would you be willing to lower your current living standard in order to curb climate change'), but responses would be difficult to compare, not knowing what was meant, exactly, by answering in the affirmative or not. If your living standard is close to or under the limit of poverty, a negative answer to this question would have another meaning than if you, according to your own judgement, live a rich and over-consuming life.

It was decided to make WWViews explicitly global in scale, the argument being that both climate change and UN negotiations about it are global, and decisions made in one part of the world will influence the lives of people in other places of the world. Citizens were expected to participate as 'world citizens', so to speak – and to address global rather than individual or local issues. Feedback from our partners indicates that citizens are perfectly capable of discussing abstract issues without having them presented in relation to local or personal experiences. This fits well with other experiences with citizen participation over the last 20 years.

Box 2.3 Main steps in the WWViews consultation

Figure 2.1 *Welcome to WWViews Day*

Figure 2.2 *Information videos*

Figure 2.3 *Deliberation in groups*

Figure 2.4 *Voting after each thematic session*

Figure 2.5 *New deliberations*

Figure 2.6 *Writing recommendations to COP15*

Figure 2.7 *Goodbye and thank you for participating*

Source: Figures: Niels Bo Bojesen

Information given

In accordance with the aim of global comparability, we wanted all information given to citizens about global warming and the climate negotiations to be identical across the participating countries. We debated with some of our partners about whether or not to include – at each national consultation – information about the position of the respective government in the climate negotiations as well as considerations about how different global political solutions might affect each individual country. However, because of the lack of comparable sources of information on climate change across the world, and because of the focus on global – in contrast to national – climate policies in the UN negotiation process, we ended by keeping nationally specific information to a minimum and focusing the consultation on global policy issues. As another measure to limit local variability, scientific experts, who are often part of deliberative exercises, were not invited to answer questions from citizens during the consultations. Our main motive for deciding against the presence of experts was the expectation that the culture of debate in several countries would encourage citizens to ask them for authoritative answers to questions that are political in nature. Thus, by making this decision, we gave higher priority to protecting deliberations from being unduly influenced by experts' opinions than to the possibility for citizens to interact with those experts.

The uniformity and the complexity of the information material given to citizens prior to the consultations risked excluding citizens from various cultural and educational backgrounds. The material had to address illiterates and academics alike, giving them an equal starting point for their deliberations, while being sufficiently detailed to provide the information needed in order to answer the quite difficult questions asked. A draft version of the information material (as well as the questions) was presented to focus groups in Bolivia, Japan, Vietnam and the US. While the feedback from these focus groups did a lot to improve the material, we also realized that additional measures could be needed in order to make it equally understandable for all citizens. One response to this challenge was that some partners decided to assemble participants prior to 26 September, since they saw a need to introduce them more thoroughly to the information material. As another measure to bring citizens closer to a common starting point for the deliberations, we decided to produce four information videos, covering the most essential information needed to take part in the deliberations. The videos were shown during the consultations as introductions to each of the four thematic sessions.

Science and bias

Besides being sufficiently detailed and understandable to citizens worldwide, the information material had to be unbiased. A journalist was commissioned to write the information material in order to help meet those requirements. The *Fourth Assessment Report* of the Intergovernmental Panel on Climate Change's (IPCC) was the main reference for the factual information given on climate change. Not only did this report encompass most relevant scientific literature, it was also widely accepted among politicians as the scientific starting point for the climate negotiations. This information was supplemented with the views of developing and developed countries and the range of economic arguments for dealing one way or another with climate change. A scientific advisory board reviewed the factual information and WWViews partners were invited to give their comments, many of which helped in avoiding possible biases.

We also sought to eliminate biases by standardizing the methodology and by providing partners with manuals on how to select citizens for the consultations and carry out the consultation meetings in detail. Considerable emphasis was placed on communicating that WWViews was not a climate campaign trying to tell people what they *should* think, but an investigation of what they *did* think. From an organizational viewpoint, one of the main criteria for selecting national partners was that they should be unbiased with regards to climate change in the eyes of the policy-makers. For example, if potential partners were considered to be too 'green', they were asked to balance the potential bias out by forming coalitions with university institutes or other organizations that were not.

Representativeness

Public participation projects are always faced with the dual challenge of including a sufficient number of citizens while doing so in an economically and practically feasible way. We make the claim that the sample of citizens consulted in WWViews is large and diverse enough to give a sense of the general trends in national and international public opinion. We do so on the grounds that all our partners have been instructed to select citizens to reflect the demographic diversity in their country or region with regards to gender, age, occupation, education and geographical zone of habitation. This was a pragmatic solution, identifying parameters easy to apply for most partners while operating with a number large enough to bring into the national deliberations a range of views that represent those most predominant in the population at large. We would argue that this is representative in the sense that the results would be very similar if the process were to be repeated with a different but equally diverse group of citizens.

From another, and even more pragmatic, point of view, the sample of citizens can be considered representative to the extent that citizens, organizers and policy-makers accept that they are. Experienced organizers know how to design their citizen consultations in a way that their politicians respect. This varies across different political contexts, which is why the strict requirement for statistical demographic selection criteria was balanced with room for national variability. In some countries, for example, ethnicity was included as a selection parameter.

While the story of WWViews is in many ways the story of reducing national variation in order to make global comparison possible, it is important to mention that decisions about homogenizing and streamlining the project were made with the under-standing and tacit acceptance that guidelines from the coordinators would not always be strictly applied. Pragmatism is an important ingredient in citizen participation, and arguably more so on a global scale.

Organizational structure and communication of results in a global context

We knew from the start that the project should be organized around many smaller na-tional meetings rather than one big global meeting. Earlier experience with European-wide citizen participation, which convened citizens from different countries in the same venue, was the extreme expense due to travel and translation costs.[1] By linking national face-to-face deliberations together through full data access and comparison over the internet, the WWViews design is much more cost-effective, thus allowing the inclusion of more countries. Once the basic project costs (establishing the global network, producing the information material, developing communication strategies, training project partners and overall project coordination) are covered, the marginal expenses (organizing a national citizen consultation and communicating the results to policy-makers and media) for including more countries are relatively low. Using the internet thus keeps the costs low while still creating a synergy between the global and the national levels and incorporating participants in a common project.

Another important reason for basing the project on national meetings was that the climate negotiations needed to be addressed on different political levels. While the lead-ers of the official policy-making process (the Danish COP Chairmanship and the UN) had to be addressed directly, it was equally important that policy-makers at the national level were informed and involved. The national level is where positions brought to the global negotiations are developed and decisions reached eventually implemented, and the national representatives at international negotiations often feel themselves primarily accountable to the citizens in their own countries. For national policy-makers, national results would often be more interesting than international ones, but by offering the pos-sibility of comparing those results at both the global and the national level, WWViews provided knowledge with the potential of linking national to global policy-making. By organizing national meetings, WWViews gained a strong national anchorage, and by making the WWViews Alliance partners responsible for communicating both national and global results to their national politicians and COP15 negotiators, they gained national ownership of the results of their deliberations. Thus, WWViews could be com-municated as both a global event and a series of national events.

WWViews had to balance the aim to make policy-makers listen to its results with the aim of being unbiased. In some ways, these two aims complemented each other but in other ways they conflicted. With no official connection to the climate negotiations, WWViews became just one in a multitude of voices competing for the attention of the policy-makers in what turned out to be the largest environmental convention in history. WWViews did have contact with the Danish COP Chairmanship and, though more sporadically, different UN institutions. While we did want their support and attention, we did not wish to sign up to their agendas such as, for example, to 'seal the deal' in Copenhagen, as this would compromise the autonomy that makes informed deliberation unique and credible as an authentic voice of ordinary people. WWViews was therefore initially received with mild interest and encouragement, as well as apprehension. Neither the COP Chairmanship nor the UN wished to subscribe to WWViews in advance, not knowing what the results would be. As WWViews proceeded (and especially as results became known), the project was more openly embraced but never integrated in the official policy-making process. Having said that, our general experience has been that WWViews has made many decision-makers acutely aware of the need to provide a space in global decision-making processes for citizen participation. Having first seen and understood what WWViews is, so it seems, people find it increasingly difficult to see how one could do without it. We have witnessed this reaction in our communications in both closed and public meetings with ministers, UN officers, politicians, ambassadors and senior political advisers.

Now that WWViews has proved itself to be not just an experiment but a viable method, the UN and other potential future initiators of WWViews-style global deliberations may be more inclined to integrate public participation in policy-making processes without knowing the results in advance and acknowledging the independence needed.

Outcomes and outlook

The aim of WWViews was to include the views of ordinary citizens in a debate that is otherwise heavily dominated by scientists, politicians and powerful interest groups. While thus offering participating citizens a chance to influence the political negotiations, it is not the task of citizen participation to replace politics, but rather to offer itself as information essential to policy-makers.

Feedback from our partners shows that the WWViews method works in diverse cultural and political contexts and is widely received among policy-makers as a credible and essential contribution to their policy-making efforts. Several partners report how news of the WWViews process and results has been positively received by politicians and decision-makers in their countries.

Although some politicians did use the WWViews results to support their political efforts leading up to COP15, and although media coverage was quite extensive in several countries, and although we did manage to present WWViews at a COP15 side event and at several other venues, the most important outcome may prove to be long term through the mere demonstration that it is possible to do such a thing in a meaningful way. WWViews has contributed to the understanding and acknowledgement worldwide of the need for increased citizen participation in political decision-making processes.

Some partners report that WWViews has sparked political interest in more citizen participation and others are already using similar methods in other contexts. WWViews has led to the establishment of a global network of organizations with an understanding of and capability to organize citizen consultations. Our hope is to activate and expand this network in the years to come and we also hope it will not be too long before the worldwide views of citizens are included in global policy deliberations on a regular basis.

Note

1 European Citizens' Consultations, www.european-citizens-consultations.eu, and Meeting of Minds, www.meetingmindseurope.org.

Part II
Understanding the Trend of International Deliberation

3

The Creation of a Global Voice for Citizens: The Case of Denmark

Annika Agger, Erling Jelsøe, Birgit Jæger and Louise Phillips[1]

The global event World Wide Views on Global Warming (WWViews) was an innovative experiment in public engagement in science and technology, aiming to create a 'global citizen voice' on climate change. This analysis is based on a study of the Danish WWViews event, drawing on theoretical perspectives of deliberative democracy and of science, technology and society (STS) studies of public engagement. The focus is on how the citizen deliberations were institutionally framed as an exercise in deliberative democracy. The analysis includes reflections on how the process was designed in order to obtain legitimacy, on how different types of knowledge and expert identities were constructed and negotiated and on how the framing impinged on the outcome. The specific conditions of the Copenhagen meeting and its relationship to a high-policy global summit (COP15) are also considered in the discussion of WWViews as an innovative design for global public engagement in science and technology.

Experiments with participatory methods of public engagement with science and technology have been carried out in Denmark at least since 1986, the year in which the Danish Board of Technology (DBT) was established. During the last two decades, public engagement exercises have proliferated across the world as part of a new mode of scientific governance, constructed in terms of a discourse of citizen engagement and dialogue between science and the public (Irwin, 2001, 2006). In this process, DBT has played a role as one of the pioneers, their participatory methods providing a source of inspiration for many public engagement initiatives. In initiating the first-ever global citizen consultation in the form of WWViews, DBT can again be seen to be acting as a pioneer. The aim of this chapter is to analyse the local WWViews event in Copenhagen as an experiment in public engagement methods on a global scale, in the light of the Danish political context and the principles of deliberative democracy on which the experiment is based.

The chapter is divided into six sections. We begin with a brief description of DBT as an institution that emerged out of a particular political context. In the second section, we outline the principles of deliberative democracy on which WWViews is based and put forward four criteria that form the basis for our empirical analysis of the event. The third section discusses the issue of the representativeness of the participants, given the foundational principle of the citizen consultation that the participants should represent the 'voice of all citizens'. The fourth and fifth sections present empirical analyses of, respectively, the ways in which the event is managed by the organizers through text and talk, and the ways in which principles of deliberative democracy play out in the citizen deliberations themselves. The analytical focus is on the ways in which principles of deliberative democracy are articulated in the interplay between top-down dynamics, involving the implementation of a standard framework for deliberations and bottom-up dynamics involving the creation of a space that opens up for a plurality of citizen voices. Finally, in the concluding section, we discuss the impact of the deliberations and reflect on the nature and significance of the Copenhagen WWViews event as an experiment in global public engagement.

With respect to methods of data analysis, we draw on the theory of deliberative democracy, analysing how deliberative principles are played out in practice at the WWViews meeting in Copenhagen. The data analysed are as follows:

- the material distributed to the citizens before and during the Copenhagen meeting;
- the guidelines formulated for organizing the event;
- the oral presentations given by the organizers and the films presented on the day;
- the collated results of participants' votes;
- sound recordings of the deliberations at three of the tables on the day (three times six hours) and transcriptions of those sound recordings;
- field notes from participant observations on the day;
- four short interviews with citizen participants on the day;
- one group interview with three citizen participants two weeks after the event; and
- one group interview with three organizers of WWViews at DBT two months after the event.

Participatory technology assessment in Denmark

The establishment of DBT in 1986 can be understood as a product of the politicization of technology assessment that began in the 1970s. This politicization was based on concerns about the consequences of new technologies and was manifested in different forms of protest against the implementation of those technologies. For example, strikes were held against the introduction of new technology in production processes and social movements emerged against nuclear power (Danielsen, 2006).

In the years that followed, concerns among workers were institutionalized into the pre-existing institutional frame of negotiations between trade unions and employees. At the end of the 1970s, these organizations reached a so-called technology agreement, giving the workers a say in decisions about the implementation of new technology

(Andersen and Jæger, 1997, p150). The idea arose for the establishment of an institution for technology assessment that could provide an institutional frame to deal with public concerns about controversial new technologies. The Danish Parliament was inspired by the Congressional Office of Technology Assessment (OTA) in the US, but instead of copying the American model they sought to construct a model building on the Danish democratic tradition (Klüver, 1995, p41).

Different scholars have identified key features of the Danish democratic tradition. One key feature relates to the notion of the 'people's enlightenment' (Cronberg, 1995; Klüver, 1995) which is informed by the thinking of the priest, poet and politician N. F. S. Grundtvig (1783–1872). Grundtvig's ideas formed the basis for the Danish 'folk high schools' in which people engage in lifelong learning, gaining the enlightenment that he considered necessary in order for people to be able to participate as responsible citizens in deliberations about society (Horst and Irwin, 2010). Transposed into the context of public engagement about science and technology, this line of thinking leads to the position that deliberation processes require the input of expert knowledge in order to provide participants with the education that allows them to exercise the rights of scientific citizenship (Irwin, 2001) and act as competent scientific citizens in participatory democratic practices, giving input to decision-making about the direction and content of developments in science and technology (Elam and Bertilsson, 2003).

Another feature of Danish democracy is its mix of representative and participatory forms (Andersen and Jæger, 1999, p333). The type of participatory democracy subscribed to draws on the thinking of the Danish theologian and teacher Hal Kock (1904–1963), who was inspired by Grundtvig. Kock argued that democracy should not be restricted to the representation of the people in political institutions such as parliament, but should also embrace participation by citizens in processes of deliberation about the common good. A third feature of the Danish democratic tradition is a strong civil society with a well-established tradition of negotiated agreements among the organizations of the labour market and a range of NGOs and social movements taking part in societal discussions.

The history of DBT and WWViews

The idea of establishing an independent institution for technology assessments was fiercely discussed in the Danish Parliament and, in the end, a small majority voted in favour of establishing DBT as an experiment (Teknologistyrelsen, 1980; Horst and Irwin, 2010). The political struggle in Parliament continued for almost ten years. In 1995 DBT was made permanent in a law stating that it was a self-governing body with the objective of monitoring technological developments, conducting independent technology assessments and communicating the results to the Danish Parliament as well as to the Danish population, with the aim of encouraging public debates on technology (Act no. 375 of 14 June 1995).[2]

Adopting a participatory approach to technology assessment, DBT experimented with methods based on principles that derive from the above understanding of the Danish democratic tradition (Teknologirådet, 1996, pp6–7). These principles are, broadly speaking, principles of deliberative democracy. The method used for WWViews is a further development of several methods developed by DBT[3] including the consensus conference (Joss and Durant, 1995), scenario workshops (Andersen and Jæger, 1999) and citizen summits (Gastil and Levine, 2005).

A number of studies have been conducted on the international adoption of DBT methods, often focusing on the question of the international applicability of the method given its emergence from a political context specific to Denmark (Joss, 1995, p91; Klüver, 1995, p45). For example, on the basis of a comparative study of consensus conferences in Denmark, Canada and Australia, Einsiedel et al (2001) conclude that it is possible to transfer the method to countries outside Europe. The authors point out that the three countries under comparison share similarities as 'post-industrial liberal democracies with common western cultural foundations' (Einsiedel et al, 2001, p94), and that it can be concluded that the consensus conference method 'travels well' to such countries. A study of the first two consensus conferences in New Zealand (Goven, 2003) showed that the international transfer of the method can be problematic and that under particular political conditions 'participatory methods may work to reinforce an already dominant expertise and the existing restrictive framing of the debate' (Goven, 2003, p437). In the case of WWViews, DBT took responsibility for the overall management of the exercise in an attempt to avoid the framing of the events in line with local political conditions.

This brief historical account of DBT draws a picture of a successful institution, but this is not the complete picture. When a right-wing government led by the Liberal Party took over in 2001, one of its first actions was to propose the closure of a range of committees, councils and advisory boards established during the former centre-left government led by the Social Democratic Party. DBT was on the list for closure but at the last second a majority in Parliament secured its survival.[4] Since then, DBT has suffered a reduction in its annual funding from the government, forcing it to seek additional funding elsewhere. Under these conditions, it is difficult for DBT to meet its remit and continue experimenting with innovative methods for including citizens in decision-making about science and technology. However, to cease being innovative would be a sure path to irrelevance. WWViews can be seen as an initiative that responds to these tensions by taking deliberation to a global level.

Critical discussions of deliberative methods

As DBT's most influential method, consensus conferences have been the subject of critical discussions. This critique belongs to the wider body of literature in science and technology studies that identifies a shift towards a new mode of scientific governance based on dialogue and citizen engagement (Irwin, 2001, 2006; Hornig-Priest, 2005;

MacNaghten et al, 2005; Wynne, 2006; Rogers-Hayden and Pidgeon, 2007; Trench, 2008). Here we focus on three main critical questions that have been raised in relation to consensus conferences as well as other forms of public engagement based on principles of deliberative democracy.

One key critical question that has been repeatedly raised is the question of the representativeness of the participating citizens (e.g. Joss, 1995, p101; Jensen, 2005, p228). If politicians have to act on the basis of the results of the conference, they want to be sure that the results reflect 'the voice of all the people'. A second, much-mentioned issue is the relationship between participating citizens and scientific experts. The interaction between experts and citizens is central to the consensus conference, with the experts being called in to contribute their specialized (and fragmented) techno-scientific knowledge, while citizens are supposed to contribute their experiential knowledge and civic accountability. But the question is whether citizens are actually capable of handling these complex issues (Irwin, 2001), and whether the different forms of knowledge are given equal weight in the deliberative process (Irwin, 2006; Jensen, 2005).

A third question concerns the framing of the issue for the dialogue (Irwin, 2001). Every consensus conference starts with the development of some information material about the issue put together by DBT and a group of experts (Andersen and Jæger, 1999). This material frames the issue, and the question can be raised about the extent to which the citizen panel is constrained by this framing and if they are free to introduce voices that articulate alternative perspectives on the issue (Goven, 2003; Jensen, 2005). It is not a question of whether or not the framing circumscribes the practices of the panel – it will always do that – but a question of the extent to which it does so. If the citizens are overly constrained by the expert framing of the issue, then the results of the conference cannot be regarded as an expression of the voice of the citizens, but as the voice of the 'framers', and thus they lose their legitimacy as an articulation of deliberative democracy.

The final question discussed in analyses of consensus conferences that we will bring up here is the question about the results and impact of the public engagement exercises (Joss and Durant, 1995; Einsiedel et al, 2001; Horst and Irwin, 2010). This question focuses on whether or not public engagement exercises actually give citizens a voice and whether this voice is heard by the decision-makers.

Deliberative democracy as a strategy for effective public engagement

In the account of deliberative democracy in this section, key premises are outlined followed by an account of four criteria that form the basis for the empirical analysis of the Copenhagen WWViews event as a public engagement exercise rooted in principles of deliberative democracy. Deliberative democracy is claimed by many political theorists to be the most influential development within contemporary democratic theory (Smith,

2009). The term 'deliberation' derives from the Latin word *de* and *libera* where *libra* refers to scale, weight or balance. Deliberation, then, involves de-scaling/weighting or balancing (Hansen, 2004, p80), and deliberative democracy is about careful consideration and dialogue as a basis for decision-making. A central assumption in deliberative democracy is that the use of political power can only be legitimate in the interest of the common good. Therefore, citizens should have the right to be heard in public processes in matters where they are affected. Another basic feature of theories of deliberative democracy is the emphasis on participation and the educational role it has for the development of citizens' democratic capabilities. It is assumed that processes of reasoning and learning take place in concrete collective practices. Following Dryzek (2000), the common denominator of the different interpretations of deliberative democracy can be defined as: individuals participating in democratic processes are amenable to changing their minds and their preferences as a result of the reflection induced by deliberation (Dryzek, 2000, p31).

Theories of deliberative democracy emphasize the role of reflection. The core idea is that preferences can be transformed in the process of deliberation, and that the participants are amenable to changing their minds when they are presented with other points of view in a non-coercive way. Deliberation is said to differ from processes of negotiation or bargaining in that deliberation rests on the merit of a decision with respect to the public good. The deliberative democracy perspective thereby rejects the perception, common to many theories of rational choice, of citizens as rationally acting individuals who seek to optimize their own particular interests. Moreover, it contests the view of many of the liberal interpretations of democracy, where citizens are regarded as having fixed interests that can be aggregated into collective decisions through devices such as voting and representation (Dryzek, 2000; Saward, 2006).

The merits of deliberative processes

The proponents of deliberative democracy emphasize four overall arguments in favour of deliberative democracy (Button and Ryfe, 2005). First, it is claimed to produce civic virtue: through education in the process of collective deliberation and the topic under deliberation, it creates more informed, active and cooperative citizens with 'better' democratic capabilities. Second, it is claimed to produce governance virtue: by stipulating fair procedures for public reasoning that are open to everyone, the outcome of deliberative procedures will be seen as politically legitimate, just and rational because they are the result of a process that is inclusive, voluntary, reasoned and equal. Deliberative processes, then, are considered to promote decisions based on the public good and to temper vested and self-interests (Hansen, 2004, p99). Third, it is argued that they produce cognitive virtue, articulating viewpoints clearly, bringing different perspectives to the issue and clarifying substantive controversies. Deliberative processes encourage participants to discover and develop their own standpoints. The exchange of arguments is governed by an ideal of creating mutual understanding and reciprocity among the participants. Fourth, deliberative democracy promotes deliberative accountability; participants in deliberative processes are encouraged to give justifications for their arguments and decisions (Gutmann and Thompson, 1996, p129).

Deliberative democracy has been subjected to critique from a post-structuralist perspective. In particular, Chantal Mouffe (for example, 2000) challenges two related assumptions central to deliberative democracy: namely, that rational decisions on political questions can be arrived at through reasoned consensus; and that consensus is the expression of 'a democracy of mankind' based on the operation of universal principles of human equality – that is, 'a consensus without exclusion' (Mouffe, 2000, p49). For Mouffe, consensus is always the expression of the hegemony of one contingent perspective over conflicting perspectives and there can never be a democracy of mankind; 'democracy can exist only for a people' (2000, p41). The notion of a democracy of mankind has a depoliticizing effect by implying that it is possible to bracket relations of power out of politics; politics is reduced to a competition between interests that can be resolved through rational argumentation (Elam and Bertilsson, 2003, p244). On the basis of this critique, Mouffe argues instead for an agonistic pluralism that acknowledges that politics is inherently conflictual and that consensus is the product of the hegemony of one perspective and not a universal rationality. We draw on this perspective in our analysis of how principles of deliberative democracy are played out in WWViews practices at the Copenhagen meeting.

Criteria for evaluating deliberative processes

Deliberative processes can be assessed in many ways, for example, in relation to the process itself and how it includes and excludes actors and the degree to which it generates civic virtue and cognitive virtue, or in relation to the output of the process and the degree to which it contributes to generating governance virtue. In the following sections we focus on the process of deliberation and, in particular, the generation of cognitive and civic virtues. At the end of this chapter we address the question of governance virtue with respect to the impact on negotiations and decision-making at COP15.

In analysing the outcome of the Copenhagen WWViews meeting, we take our starting point in Graham Smith's (2009) framework for analysing the degree to which democratic innovations realize four explicitly democratic goods – namely, inclusiveness, popular control, considered judgement and transparency. It is argued that these criteria are fundamental to almost any theoretical account of the democratic legitimacy of institutions (Smith, 2009, p12). We have chosen to use three of these criteria as a basis for analysis: inclusiveness, popular control and considered judgement, reframing them slightly.

Inclusiveness draws attention to the way in which political equality is realized in relation to presence and voice. Applying this criterion, we focus on the issue of representativeness with respect to the selection of participants and the question of who participates in practice. *Popular control* deals with the degree to which participants can influence different aspects of the deliberative process. We address this criterion by looking at how the principles of deliberative democracy are at work in the interplay between top-down and bottom-up dynamics in the organizers' framing of the event. *Considered judgement* involves enquiry into citizens' understanding of the issue under consideration and the perspectives of the other participating citizens. Here, we focus on the citizen deliberations themselves and the degree to which they open up for a plurality of different voices and for dialogue across those voices. Our position is not that dominance of some claims over others and some voices over others is a problem in itself – indeed it is inevitable – but that it is important to acknowledge the inevitable relations of dominance and subordination and to engage in detailed empirical analysis of those relations with a view to producing insights that can further future participatory practices.

We do not view democratic assessment simply as a question of judging whether an exercise in deliberative democracy is 'democratic' as opposed to 'non-democratic' (Agger and Löfgren, 2008). Democracy is a matter of 'more or less', rather than a Pareto-optimal state of governance that may be attained. Thus we treat each criterion as a continuum in which the questions forming the basis of evaluation are phrased in relative terms – to what extent or to what degree is the criterion fulfilled (Beetham, 1993, p3).

Representativeness

In public engagement exercises, the participants are typically understood as ordinary citizens and expected to represent the voice of the citizens. How the voice of the citizens is defined is, however, a complex and vexing question. DBT has been asked the question since the very first consensus conference was held in 1987; their answer is that they do not aim for representativeness (Klüver, 1995, p46). Even in a small and relatively homogeneous country like Denmark, small groups of citizens (in consensus conferences we are dealing with a panel of 14–16 citizens) cannot ensure a statistically representative sample. Instead DBT aims for a panel composed of multiple voices. As the Director of DBT, Lars Klüver describes it:

> *Representativeness of the panel is not relevant ... The consensus conference method is based on dialogue between people with different backgrounds and different sets of values ... Therefore, we should ensure that the panel should reflect as many different views as possible – representation cannot ensure that, but a demographically 'mixed' lay panel can; providing a feasible, relevant and transparent solution.*
> (Klüver, 1995, p46)

Irwin (2001) points to the difficulty of applying the Danish understanding of representativeness within the British context of public engagement in a different political culture. In a public consultation on developments in the biosciences in the late 1990s, for example, the political intention was to design an exercise that was 'citizen-led and participatory'. During the planning of the exercise it was decided that the qualitative input (based on two-day workshops held in six different places involving 120 citizens) should be supplemented by a quantitative input of interviews with 1100 members of a representative people's panel. In this way, the consultation 'moved from its initial unformed and open stage ... to a large-scale and "representative" exercise based upon a sophisticated social research methodology' (Irwin, 2001, p8). This move was, among other things, '... in order for the study to be taken seriously by ministers and other observers' (Irwin, 2001, p8).

Thus an assumption shaping the organization of the British case was that the exercise would only be taken seriously by politicians if it were based on a representative panel of citizens. This assumption builds on the principles of representative as opposed to deliberative democracy. From the point of view of representative democracy, public participation exercises lose their legitimacy and their results risk being ignored if the panel is not statistically representative of all citizens. This understanding of representativeness underpins many of the questions that DBT frequently faces: Haven't you just given a voice to a group of white, elderly, well-educated males? In other words, what about relatively weak citizens? How can we be sure that you have not recruited an interest group or people with a particular political attitude?

Representativeness by selection

DBT has responded to the question of the under-representation of relatively weak citizens through reference to the selection procedure. In the case of the WWViews event, DBT selected 3500 citizens at random from the national register of citizens and sent them invitations to participate. Approximately 500 citizens responded in the affirmative. Of them 100 were then selected on the basis of criteria of age, sex, education, occupation and geographical location in order to gain the widest socio-demographic spread,[5] and on the day itself 97 citizens showed up. In this way the group of participating citizens were socio-demographically 'mixed' and thus it can be argued that DBT fulfilled their aim of including many different voices.

Nevertheless, a careful selection of participants cannot ensure that *every* group in society has a voice in the exercise. DBT has not been able to find an answer to the question about the under-representation of minority groups. They acknowledge that under-represented groups in Danish society such as poor people, the homeless and some ethnic minorities are not included in the citizen panel because they decline the invitation. But this is not a problem unique to public engagement exercises; these groups of citizens are also difficult to reach in the case of other methodologies such as surveys. It is also a general experience that some ethnic minorities are very difficult to mobilize in other kinds of political processes (Crowley, 2001), and it will probably demand other methods to recruit these groups. On this basis, giving voice to *all* groups in society must be regarded as an unfulfilled ideal in exercises in deliberative democracy, just as it is in institutions of representative democracy such as parliament.

The question about the over-representation of people with particular interests at stake deals with the political attitudes and values of the participants. Very early on, DBT was confronted with this question and its underlying assumption that, for example, recruitment among environmental activists represents a threat to the legitimacy of the results of the exercise. DBT tried to answer this question by a study of one of the first consensus conferences in 1989, commissioning two psychologists to carry out a study of the composition of the citizen panel and their motivations for participating. The conclusion of the study was that there were no signs of citizens 'with hidden agendas about getting organized interests represented in the panel' (Rienecker and Erichsen, 1990, p12).[6]

Indeed, it could be argued that if the participating citizens are supposed to represent the population, some of the participants ought to be activists since they represent a group in society and thus deserve a voice in the exercise. Moreover, the figure of the ordinary citizen and the idea of the singular, unified voice of the citizens can themselves be criticized for romanticizing the public and presenting them as a homogeneous, unitary entity (e.g. Horst, 2008, p262).

Representativeness by values

The idea that participating citizens ought to represent the political attitudes and values of the population as a whole is based on the notion of political representativeness. Until now, DBT have refused to claim representativeness in the traditional political sense of the concept, with the argument that the only group of people in Denmark that can claim to be representative from a political standpoint is the Danish Parliament (Klüver, 1995, p46).

According to traditional understanding of political representativeness (e.g. Pitkin, 1967), groups in society have fixed interests and identities, which the elected person has to represent in the political institution on behalf of the rest of the group. However, this understanding has been challenged in newer perspectives in political science. Here it is argued that, 'there is no self-presenting subject whose essential character and desires and interests are transparent, beyond representation, evident enough to be "predicted by" their appearance or their behavior' (Saward, 2006, p312).

Saward's (2006) point is that representativeness always entails somebody claiming to represent somebody or something to an audience; if the intended audience does not hear the claim, or is not convinced by the claim, then there is no representation. This is actually what the two above-mentioned cases from Denmark and the UK show. In the Danish case, the politicians (the audience of the claim), probably due to the Danish political culture, seem to accept that the citizen panel represents the citizens (as DBT claims) and find the results legitimate. In the British case, the politicians are not convinced that a qualitative citizen-led exercise represents the voice of the citizens and so this claim does not work to the same extent but has to be supplemented with a quantitative survey with a statistically representative citizen sample.

In the process of convincing the audience about the representative claim, a 'mutual constitution' (Saward, 2006, p314) of the representative and the represented group takes place. It is in this process that the interests, attitudes and identities of the represented are shaped along with the acceptance of the representative claim. This is what DBT claims is going on in deliberation exercises. Through the process of deliberating, the participants' attitudes and values about the issue in question are not only expressed but also shaped and transformed. Due to the well-informed and reasoned nature of the deliberations and the demographic mix of the participants, which bring different experiences and values into the deliberation, the results of the deliberation are supposed to resemble results of a similar exercise with another panel of demographically mixed citizens. In this way, the results of the exercise can be understood as representative.

Framing the deliberations through official texts and talks

In this section we address the ways in which the event is managed by the organizers through text and talk, including information material, lists of questions for deliberation and speeches. The focus is on how principles of deliberative democracy are at work in the interplay between top-down dynamics, entailing the application of a fixed framework, and bottom-up dynamics, entailing the creation of a site for the articulation of a plurality of citizen voices.

In WWViews, the text and talk were designed in advance in order to create a fixed and standard framework for specific activities. In that sense, the management was top-down. The citizen deliberations were structured in terms of four sessions in which tables of six citizens (each with a facilitator) engaged in deliberations related to a set of specific questions and, at the end of each of the sessions, voted individually on their responses to those questions. Following these four sessions, a final session was held, oriented towards consensus formation, the task being for each table of participants to formulate a common message for negotiators at COP15 in the form of their most important recommendation in the light of the day's deliberations. In pre-formulating the questions for deliberation, in order to tie the event closely to the COP15 process and optimize its impact,[7] the organizers diverged here from their usual practice in participatory events of allowing the participants to formulate the agenda themselves. The deliberations were also framed through the distribution of a set of ground rules for good dialogue that was placed at each table prior to the deliberations:

> *Speak openly and honestly – say what you mean; listen to what the others are saying; show respect for everyone; contributions should be brief and precise; focus on the topic; the facilitator decides on the order of speakers; the facilitator does not participate in the debate.*

Principles of deliberative democracy are clearly at work here: the guiding precepts are of a purely procedural nature, designed to ensure that there are no illegitimate constraints on the individual's participation in dialogue, and the use of the imperative – positioning the participants as under an obligation to act in the way stated – indicates that good dialogue depends on obedient adherence to a pre-set procedure.

But, while the framework is tightly managed in order to achieve specific goals and output, bottom-up dynamics are also at play, since the information materials, speeches and ground rules have been designed on the basis of principles of deliberative democracy, in order to provide a space for well-informed and active participation on the part of the citizens. A central aim of deliberative exercises, as noted above, is to empower citizens with the rights of scientific citizenship within a context that supplies the political education that, according to this way of thinking, allows them to exercise those rights responsibly (Elam and Bertilsson, 2003, p242).

The aim of our analysis is to open up for reflexive consideration how deliberative principles in practice are evident in the interplay between top-down and bottom-up dynamics. Our assumption is that such reflexivity can produce insights to inform the practice of public engagement with science. There is space here only for an illustrative analysis of the information booklet that was distributed to participants in advance of the meeting and on the day.[8]

Figure 3.1 *The WWViews deliberations in Denmark*

Source: Photo: Jørgen Madsen, DBT

The information booklet

The information booklet presents expert knowledge about global warming as input to the deliberations, in line with the principle intrinsic to the model of deliberative democracy that citizens' judgements should be well-informed. A great many scientific knowledge claims are put forward as categorical statements of established facts. At the same time – and in line with the aim of providing a well-balanced picture – limits of scientific knowledge are also acknowledged, with lack of knowledge being attributed not to scientists but to a vague, collective 'we': 'We know nothing about how much heat is needed to trigger this process. It may be happening right now. Nor do we know how fast it will be. It may take several hundred years' (p14).

Throughout, the perspective of the Intergovernmental Panel on Climate Change (IPCC) dominates. This weighting is in line with the declared editorial line of the booklet: 'This paper largely builds on the latest assessment report from the panel, published in 2007' (p3). And it is stated explicitly that this panel is 'the authoritative source for this kind of knowledge' (p3). In line with the principle of deliberative democracy that those judgements should result from a weighing up of truth claims, competing views about the problem of global warming and solutions to the problem are presented, but these views are given less weight than IPCC discourse. For instance, the sceptical perspective represented by Bjørn Lomborg is not integrated into the text itself, but is instead presented in a separate box. The near-consensus status of the IPCC discourse is implied through a backgrounding of the terrain of discursive struggle in which the IPCC discourse competes with alternative discourses. Differences between points of view are repeatedly presented as differences between individuals' opinions, detached from the wider political context in which competing discourses clash and opposing interests are at stake. Take the following example:

> There are 'climate change deniers' who claim that global warming doesn't have any
> basis in reality. Others think that climate change is a reality but that it is not due to
> human-made greenhouse gases. So they do not think that a new climate agreement
> is necessary or relevant. (pp20–21)

Moreover, truth claims belonging to the IPCC discourse are given strong support through various rhetorical features. For example, the point that global warming is largely due to human-made greenhouse gases is given backing through an appeal to scientific consensus: 'Most of the global warming observed since 1950 has been due to human-made greenhouse gases. Scientists are now at least 90 per cent certain that this is the case' (p10). And, with respect to the political process associated with the Climate Convention and the Kyoto Protocol, political consensus is invoked: 'The vast majority of countries have found that the achieved results are not sufficient compared to the challenges. They decided in 2007 in Bali that a new climate agreement should be made' (p20). At the same time, the booklet is presented as if it were neutral: it is stated in the foreword of the Danish version that the information is objective (while the English version uses the more moderate term, well-balanced, p3), rather than the product of a political position-taking with respect to the terrain of discourse on global warming.

The framing of the issue in terms of the discourses of IPCC provides participants at the meeting with potential resources for debates about the topic. On the other hand, it also may work to circumscribe the citizen deliberations in a way that leads to the exclusion or marginalization of non-IPCC discourses, representing alternative ways of talking about, and therefore understanding, global warming. This is not necessarily a problem in itself – and the privileging of IPCC discourses may indeed be based on a reasonable evaluation of the value of the alternatives. But it is a problem that, through the reduction of political struggle to disagreement between individuals' views and the presentation of the booklet as neutral, the contested nature of the political terrain is consigned to the background, and IPCC's positioning within that contested political terrain is underplayed. As Blok (2007, p176) has pointed out, deliberative discourse, with its ideal of unconstrained, reasoned dialogue, fails to acknowledge the framing and selection processes at work in the use of expert knowledge as a basis for rational argumentation. And, as noted earlier in the chapter, this is problematic, judged from the perspective of a post-structuralist model of democracy that contends that politics is intrinsically conflictual, and consensus always the expression of the hegemony of an (always contingent) discourse (Mouffe, 2000, p49).

Enacting principles of deliberative democracy in the citizen deliberations

In this section, we turn to how the principles of deliberative democracy are played out in the practices of the citizen deliberations. Empirical analyses of the processes of interaction in public consultation practices are actually relatively rare, the data often being restricted to the written texts produced (information material and the reports written by participants) and interviews with organizers and, in a few cases, citizen participants. Our account has two themes. The first concerns relations between top-down and bottom-up dynamics, addressing the analytical question: What part do facilitators play in managing the tension between creating space for a plurality of citizen voices and steering the process in order to achieve a form of closure in line with pre-formulated aims? The second theme concerns the nature of the deliberations as spaces in which expertise is democratized in the sense that citizen participants can lay claim to expertise, based, for instance, on experiential knowledge forms. The focus here is on how those spaces are politicized such that claims to expertise are subject to social contestation, with some knowledge claims being given more weight than others by the participants and some participants being positioned as more authoritative than others; deliberative democracy does not entail a level playing field, although the deliberative model aspires to that ideal.

Figure 3.2 *Deliberations at the tables*

Source: Photo: Jørgen Madsen, DBT

As noted earlier, our position is that the dominance of some claims over others and some voices over others is inevitable and not itself a problem but that it is important to engage in detailed empirical analysis of the hierarchy of voices with a view to producing insights that can further future participatory practices. Here, there is no space to present the detailed analysis we argue for; we have space just to give a few illustrative examples of the role of the facilitators in managing the tension between opening up for citizen voices and steering the discussion in the intended direction, and of how the participants negotiate between different knowledge claims and (expert) identities.

The role of the facilitators

The facilitators managed the interaction at their tables in adherence with principles of deliberative democracy through a number of different strategies, applied deliberately or not. One strategy was to summarize the ongoing discussion and to present it in terms of an opposition between two different positions. This characterization of the discussion in terms of two opposing positions is obviously in line with the deliberative principle that participants should form a considered judgement by weighing up different perspectives. It can be criticized as a reduction of complexity that narrows down the discursive field to fit the terms of the facilitator's own interpretative lens and thus constrains what can be said. Alternatively, it can be understood as a constructive move that helps the participants to position themselves in the debate and take a clear stance.

Another strategy in play was the point that views were open to change through the meeting of different perspectives in deliberation with others:

> *Jørgen: I also think that our first discussion about the degree of concern means that all of us around the table would say that something has to happen, so that's in line with the first theme.*
>
> *Facilitator: And what about the last two themes?*
>
> *Jørgen: Yes, well, now I was one of those who answered 'a bit concerned', so for me it would suit me fine with two degrees. On the basis of the material I've seen and the discussion, that would be the level that I would vote for. Now I'm going to hear the debate.*
>
> *Facilitator: Yes, you may well arrive at something else.* (Table recording from table 1, lines 53–63)

Here, the participant, Jørgen, makes the meta-comment that the debate so far has indicated consensus round the table. He asserts the position he has arrived at in the debate, partly as a result of input from the information material, but also indicates that his own position may change as a result of the debate to follow ('now I'm going to hear the debate'). The facilitator makes this openness explicit in stating, 'Yes, you may well arrive at something else.' Thus both the participant and the facilitator express the ideal of deliberative democratic debate: that participants should be open to the different arguments presented in the information material (designed to provide an informed starting point for discussion) and in the discussion itself, and then form a final position on the basis of hearing those arguments.

The democratization of expertise in practice

As expertise is democratized, a negotiation between different knowledge claims and identities takes place and a tension often occurs between an opening up to a plurality of voices, on the one hand, and a move towards consensus formation, on the other, as in the following example:

> *Jakob: It's also a question of putting on the pressure so that we can get developments that mean that we save fuel.*
>
> *Peter: That's correct but you can't change fuel from A to B with a lorry.*
>
> *Christina: But what about something like ethanol, does that not pollute less?*
>
> *Peter: No, it doesn't pollute less.*
>
> *Christina: It doesn't?*

Peter: That's because you can say, well, if it is bio-ethanol, then it's CO_2 neutral [...] but making it isn't cheap.

Christina: No, not cheap, but isn't it CO_2 neutral? It doesn't produce so much CO_2.

Peter: Yes, it does at first, but the green plant you plant, it absorbs it so it evens out. That's why it's called CO_2 neutral. (Table recording from table 1, lines 732–751)

Peter positions himself as an authority in relation to ethanol, but Christina does not capitulate. First, Christina challenges Peter again:

Christina: It's the whole system we should look at.

Peter appeals to an implied consensus:

Peter: No, but it's the same as what happens with the cost of moving a good from A to B, it goes up. We don't want that, do we?

Christina, in her turn, again challenges the consensus status of Peter's claim:

Christina: Well actually, I don't know if we don't want that.

Peter recognizes Christina's position ('Well, but well okay') and then argues against it:

Peter: Well, but well okay. It's the same as putting the price of oil up. Well, oil gets dearer ...

Christina meets this argument with a counter-argument:

Christina: Yes, yes. But the price difference won't be so big if we raise the price of fossil fuels.

Jakob backs up Christina's position:

Jakob: Well, it's the fossil fuels you should raise the price of in order to encourage people to –

Christina: Use something else

Peter challenges Christina's position and is then backed up by Anna:

Peter: But that's what I'm getting at, it'll also become dearer to move from A to B, because it's more expensive today to produce that bio-ethanol.

Anna: And that's exactly what I meant when I said that somehow or another, it's got to be sustainable in some way or another, because it's something we exploit. (Table recording from table 1 lines 753–775)

These examples illustrate how deliberative interactions are dynamic and expert authority is not something that is fixed and tied to particular individuals, but subject to social processes of negotiation. While one participant, Christina, can be said to be ascribed, and to ascribe herself, a subordinate position with less authority, there is space here for her to challenge this and to be heard by the others. This points to a relatively open process, in line with the principle of deliberative democracy, that deliberations should make room for a plurality of voices, although there may be no symmetry of, or equality between, the voices.

Another requirement of processes of deliberative democracy is that everyone has the right to question the designated topics of discussion and to initiate reflexive arguments about the procedural rules (Benhabib, 1994, p31). The following is an example of such questioning and reflexivity on the part of a participant. She expresses frustration based on reflexive recognition of a process in which the complexities of the issue are set aside in the move towards formulating a single plan for action:

> *Susanne: Well, I'm sitting here thinking a bit about, well the statement we have to come up with at some point and it has to be max 30 words. And if I now could imagine this responsible politician who really takes this seriously, note that I'm saying 'responsible politician', what is needed in order to influence this responsible politician? Well, does it just have to be a white piece of paper with 30 words or should we fold it like a paper airplane and make a sum? What is needed exactly to say to this politician, this is what we've got to do ... What should the form be? Well, 30 words, yes. But a long sentence has to be formulated which may tend to be very general, so you read, 'yeah, we have to stop global warming, we have to have alternative energy, we have to have this and that'. Well, primary six could also come up with that.*
>
> *Marie: Yes, they could.*
>
> *Facilitator: Well, Connie Hedegaard asked for a signal for her negotiations, what she's got a mandate for. What will we as citizens accept?* (Table recording from table 2, lines 676–714)

Susanne's reflexive critique of the shift from deliberative consideration of the complexities of the issue to a discussion of how to communicate a simple, singular message is met by the facilitator's attempt to justify the exercise – restating the purpose of WWViews, whereby citizens are treated as central agents in the participatory political process: 'Well, Connie Hedegaard asked for a signal for her negotiations, what she has got a mandate for. What will we as citizens accept.' Also about five minutes later, the facilitator reiterates the need for simplicity on pragmatic grounds: 'Yes, that's really the way we'll get the most messages across in practice.' There were several other instances in which participants put forward reflexive, meta-comments about the design of the debate, including raising questions about the formulations of the questions. In all cases, they were met by explanations by the facilitators, defending the design on the grounds that it furthered the pragmatic aim of sending a message to COP15 that would have an impact on the negotiations.

Discussion and conclusion

The above empirical analyses focused on how deliberative principles are enacted through the framing of the deliberations in text and talk, through the facilitators' management of the process and through the participants' negotiations of expertise. There was space for only a few examples, but those examples indicate the complexity of the practices – a complexity which means that the processes can neither be described as a simple reproduction of deliberative principles nor dismissed as an empty exercise in the legitimization of scientific policy-making through the enlistment of citizens in a process that is only bottom-up in its rhetoric of public participation and dialogue.

The basic idea behind deliberative processes like WWViews is to extend democracy and give citizens a voice in discussions about complex issues involving science and technology. This raises the question about whether their voice is heard, and to what extent their views and recommendations have an impact on actual decisions taken by politicians or other decision-makers in the field that the citizens' deliberations address.

Expectations of this kind were also formulated by the Danish Minister for Climate and Energy, Connie Hedegaard, who was an ambassador for WWViews, when she said at the opening of the Danish WWViews event on 26 September:

Figure 3.3 *The Danish Minister of Climate and Energy, Connie Hedegaard, at the Danish WWViews event*

Source: Photo: Jørgen Madsen, DBT

'Do help us politicians in the next months so that we can help the globe ... There is a need for support for an ambitious agreement, there is a need for you to create transition, and then, not least, there is a need for you to secure that the political price for not acting and not taking the necessary decisions here in Copenhagen in December, it will be so high that no government can afford to pay that price.'
(Hedegaard, 2009)[9]

Similarly, many of the citizens who participated in the Danish WWViews event expected to be heard. This was expressed in short interviews we made just after the end of the event on 26 September. One citizen said, when asked whether he expected that their recommendations would influence the outcome of COP15, 'Yes it probably can, if it is fairly unambiguous from most of the world, then I absolutely believe that it will have an effect'; and another said, 'Yes' and continued '... well, because I think that ... that something must be done, and I think that we are coming up with the right recommendations.' Yet another citizen answered the same question, 'Not really, but then at least we have a clear conscience. We have done what we could.' So even though the majority expected that the recommendations would have an impact, some doubt was also expressed.

More generally, if we consider the experience from the Danish consensus conferences, there are only a few examples of direct 'measurable' impacts on political decisions as a result,[10] despite the fact that DTB has almost optimal conditions for gaining access to politicians because of its close relationship with the Danish Parliament.[11] When discussing the impact of both consensus conferences and other deliberative processes, the focus on such concrete political outcomes is too narrow, however. These events may have more indirect impacts on the broader political situation and agenda within a field, and they may influence the way deliberative procedures are viewed by dominant actors and may lead to inclusion of deliberative practices in political processes. Furthermore they have a symbolic value as a demonstration of the ability of lay people to take part in complex decision-making (Einsiedel et al, 2001).

From this broader perspective it is clear that a deliberative process may have social and political impacts beyond its direct political consequences. Nevertheless, it is also clear that many of the members of the citizens' panel expected the results of WWViews to directly influence the agreements at COP15. Moreover, such an influence or lack of influence would be easy to detect in the case of WWViews, since COP15 took place only a few months later under intense media attention. Despite this, it is also clear that the recommendations of WWViews had little influence on the outcome of COP15.

WWViews can be said to have given agency to participants by treating them as scientific citizens and opening up for a plurality of citizen voices in the deliberations themselves. In treating participants as scientific citizens, the practices may well have contributed to creating more informed and active citizens – in the terms of deliberative democracy, the practices had civic virtue. But, at the same time, the lack of direct impact on COP15 can be seen as a challenge to the raison d'être of WWViews. The overarching purpose was to demonstrate that political decision-making processes on a global scale can benefit from the participation of ordinary people.[12] On the other hand, the absence of a direct impact is not new in the history of deliberative exercises. It is nevertheless a challenge to these exercises, which reflects the institutional tension between the dominant notions and practices of representative democracy and the, until now, much weaker and more fragile experiments with deliberative procedures. This is, of course, particularly true in the case of an international and politically highly profiled and complex event such as COP15.

With respect to the other categories of social and political impact mentioned above, no systematic studies have been made so far and it is probably too early to make an assessment. As regards the influence on public debate, the press coverage of WWViews in Denmark was not extensive, although it did include an almost five-minute feature on prime-time television news on the Danish Broadcasting Corporation on 27 September 2009. However, the overall positive reception by many different stakeholders in Denmark and internationally seems to indicate that even though WWViews did not gain much direct political influence, the initiative may have an impact on the international recognition of deliberative procedures as policy tools. There is no doubt that such an effect was also one of the aims of DBT; and in many countries WWViews also functioned as a democracy-building project, which influenced ideas about democracy, including issues such as women's participation in democratic procedures.[13]

Thus, WWViews may have influenced the broader political climate in civil society, even on an international scale. In the longer term this may be an important outcome of WWViews. Undoubtedly, it has also contributed to the already increasing attention to, and interest in, deliberative processes internationally. The simultaneous expression of a citizen voice in 38 countries, and the unambiguous appeal for action in relation to climate problems from so many different groups of citizens seems impressive despite the apparent lack of influence. The general political momentum that drives the interest in deliberative processes (i.e. crises of governance and political legitimacy) and the positive reception of WWViews by many stakeholders and decision-makers also point in this direction. While public engagement exercises are still marginal, even in countries like Denmark where they are institutionally well established, WWViews may serve to promote public engagement in countries where it is less well established.

WWViews has a different approach from consensus conferences in some respects, particularly regarding the role of expertise in the process. The lack of direct dialogue with experts meant that expert knowledge was only expressed via the material that was produced in advance, that is, the information booklet and videos. As we tried to show in our analysis, this material largely canonized IPCC as the only real and reliable expertise. Perhaps as a consequence of this, the dialogue between the citizens in several situations led to attempts to establish expert positions within the groups that could not be dealt with at the tables through dialogue with experts. There is a need to consider this issue in the development of the approach, for instance through a more transparent and reflexive approach to the presentation of different views in the information material (which does not necessarily allow all conflicting views to be presented as equally good or valid).

The deliberative procedures analysed above are based on an ideal of participation by citizens, free of organizational or other interests that presumably could distort the deliberative process. As already mentioned, organizational interests are a legitimate part of any modern society, and public involvement up to and during COP15 was comprised of a whole range of different popular activities initiated by NGOs and local groups of grass-roots activists. Indeed, the results of WWViews were presented at the unofficial Climate Forum in Copenhagen during COP15, which was the most extensive representation of Danish and international NGOs. So the conceptual (as well as practical) division between NGOs and other organized groups in civil society, and the 'purified' community of 'ordinary citizens' in the deliberative process, can be questioned. It could also be argued that deliberative activities should be connected more openly to other public activities, including forms of social activism, focusing on controversial issues such as climate change.

Notes

1 The authors were not involved in the planning and organization of WWViews and have no formal relationship with the Danish Board of Technology (DBT), the Danish organizer of WWViews. The research was based on a cooperative agreement between the authors and DBT. This agreement included permission for the authors to carry out participant observation at the WWViews event in Copenhagen Town Hall, access to related materials, interviews with DBT organizers of WWViews and the authors' presentation of research results to the organizers for comments.

2 For other accounts of the history of DBT, see for example, Klüver, 1995; Andersen and Jæger, 1999; Horst and Irwin, 2010.

3 Interview with DBT project organizers (director of the secretariat, head project manager and chairman on the WWViews day in Copenhagen), 24 November 2009.

4 Interview with DBT organizers, 24 November 2009.

5 Interview with DBT organizers, 24 November 2009.

6 Translated from the Danish by the authors.

7 Interview with organizers of the Danish WWViews, 24 November 2009, lines 2148–2173.

8 Unless otherwise noted, all citations are the authors' translation of the Danish-language edition of the information booklet. We have chosen this option rather than using the English-language edition, since it is the Danish-language version that was in use at the meeting in Copenhagen.

9 Authors' translation.

10 One example, which has been mentioned repeatedly, dates back to the first Danish consensus conference on gene technology in agriculture and industry in 1987. In the final document the citizens proposed a ban on gene technology in animals. As a consequence, funding of gene technology research projects in animals was abandoned in the first Biotechnological Research and Development Programme passed by Parliament shortly afterwards.

11 This is not only a formal institutional relationship but also a relationship articulated in practice. Thus, when a final document from a Danish consensus conference is presented on the last day of the conference, members of the relevant parliamentary committee are present and comment on the recommendations and discuss with the citizen panel.
12 From the WWViews website: www.wwviews.org.
13 Interview with DBT staff, 24 November 2009.

References

Agger, A. and Löfgren, K. (2008) 'Democratic assessment of collaborative planning processes', *Planning Theory*, vol 7, no 2, pp145–164

Andersen, I.-E. and Jæger, B. (1997) 'Involving citizens in assessment and the public debate on information technology', in A. Feenberg, T. H. Nielsen and L. Winner (eds) *Technology and Democracy: Technology in the Public Sphere: Proceedings from Workshop 1*, University of Oslo, Oslo

Andersen, I.-E. and Jæger, B. (1999) 'Scenario workshops and consensus conferences: Towards more democratic decision-making', *Science and Public Policy*, vol 26, no 5, pp331–340

Beetham, D. (1993) *Auditing Democracy in Britain: Democratic Audit*, University of Essex, Scarman Trust

Benhabib, S. (1994) 'Deliberative rationality and models of democratic legitimacy', *Constellations*, vol 1, no 1, pp25–53

Blok, A. (2007) 'Experts on public trial: On democratizing expertise through a Danish consensus conference', *Public Understanding of Science*, vol 16, no 2, pp163–182

Button, M. and Ryfe, S. M. (2005) 'What can we learn from the practice of deliberative democracy?', in J. J. Gastil and P. Levine (eds) *The Deliberative Democracy Handbook: Strategies for Effective Civic Engagement in the 21st Century*, Jossey-Bass, San Francisco, CA

Cronberg, T. (1995) 'Do marginal voices shape technology?', in S. Joss and J. Durant (eds) *Public participation in science: the role of consensus conferences in Europe* London, Science Museum with the support of the European Commission Directorate General XII, pp125–134

Crowley, J. (2001) 'The political participation of ethnic minorities', *International Political Science Review*, vol 22, no 1, pp99–121

Danielsen, O. (2006) *Atomkraften under pres: dansk debat om atomkraft 1974–85*, Roskilde University Press, Roskilde

Dryzek, J. S. (2000) *Deliberative Democracy and Beyond – Liberals, Critics, Contestations*, Oxford University Press, Oxford

Einsiedel, F., Jelsøe, E. and Breck, T. (2001) 'Publics at the technology table: The consensus conference in Denmark, Canada, and Australia', *Public Understanding of Science*, vol 10, no 1, pp83–98

Elam, M. and Bertilsson, M. (2003) 'Consuming, engaging and confronting science: The emerging dimensions of scientific citizenship', *European Journal of Social Theory*, vol 6, no 2, pp233–251

Gastil, J. and Levine, P. (eds) (2005) *The Deliberative Democracy Handbook: Strategies for Effective Civic Engagement in the 21st Century*, Jossey-Bass, San Francisco, CA

Goven, J. (2003) 'Developing the consensus conference in New Zealand: Democracy and de-problematization', *Public Understanding of Science*, vol 12, no 4, pp423–444

Gutmann, A. and Thompson, D. (1996) *Democracy and Disagreement*, The Belknap Press of Harvard University Press, Cambridge, MA

Hansen, K. M. (2004) *Deliberative Democracy and Opinion Formation*, University Press of Southern Denmark, www.kaspermhansen.eu/Work/Hansen2004.pdf

Hedegaard, C. (2009) *Opening Speech at the Danish World Wide Views event on the 26th of September 2009*, Copenhagen, www.youtube.com/watch?v=_sMX5iajGYc&feature=related, accessed 11 February 2010

Hornig-Priest, S. (2005) 'Commentary: Room at the bottom of Pandora's box: Peril and promise in communicating nanotechnology', *Science Communication*, vol 27, no 2, pp292–299

Horst, M. (2008) 'In search of dialogue: Staging science communication in consensus conferences', in D. Cheng, M. Claessens and N. R. J. Gascoigne et al (eds) *Communicating Science in Social Contexts: New Models, New Practices*, Springer Publishing, New York

Horst, M. and Irwin, A. (2010) 'Nations at ease with radical knowledge: On consensus, consensusing and false consensusness', *Social Studies of Science*, vol 40, no 1, pp105–126

Irwin, A. (2001) 'Constructing the scientific citizen: Science and democracy in the biosciences', *Public Understanding of Science*, vol 10, no 1, pp1–18

Irwin, A. (2006) 'The politics of talk: Coming to terms with the "new" scientific governance', *Social Studies of Science*, vol 36, no 2, pp299–320

Jensen, C. B. (2005) 'Citizen projects and consensus-building at the Danish Board of Technology: On experiments in democracy', *Acta Sociologica*, vol 48, no 3, pp221–235

Joss, S. (1995) 'Evaluating consensus conferences', in S. Joss and J. Durant (eds) *Public Participation in Science: The Role of Consensus Conferences in Europe*, Science Museum with the support of the European Commission Directorate General XII, London

Joss, S. and Durant, J. (eds) (1995) *Public Participation in Science: The Role of Consensus Conferences in Europe*, Science Museum with the support of the European Commission Directorate General XII, London

Klüver, L. (1995) 'Consensus conferences at the Danish Board of Technology', in S. Joss and J. Durant (eds) *Public Participation in Science: The Role of Consensus Conferences in Europe*, Science Museum with the support of the European Commission Directorate General XII, London

MacNaghten, P., Kearnes, M. and Wynne, B. (2005) 'Nanotechnology, governance and public deliberation: What role for the social sciences?', *Science Communication*, vol 27, no 2, pp268–291

Mouffe, C. (2000) *The Democratic Paradox*, Verso, London

Pitkin, H. F. (1967) *The Concept of Representation*, University of California Press, Berkeley, CA

Rienecker, L. and Erichsen, F. (1990) *Lægfolk i en konsensuskonference [Lay-people in a Consensus Conference]*, The Danish Board of Technology, Copenhagen

Rogers-Hayden, T. and Pidgeon, N. (2007) 'Moving engagement "upstream"? Nanotechnologies and Royal Society and Royal Academy of Engineering's inquiry', *Public Understanding of Science,* vol 16, no 3, pp345–364

Saward, M. (2006) 'The representative claim', *Contemporary Political Theory*, vol 5, no 3, pp297–318

Smith, G. (2009) *Democratic Innovations Designing Institutions for Citizen Participation*, Cambridge University Press, Cambridge

Teknologirådet (1996) *Ti år med Teknologinævnet. En tiårsberetning om Teknologinævnet 1986–1995*, Teknologirådet, Copenhagen

Teknologistyrelsen (1980) *Teknologivurdering i Danmark – betænkning afgivet af et udvalg under Teknologirådet*, Teknologistyrelsen, Copenhagen

Trench, B. (2008) 'Towards an analytical framework of science communication models', in D. Cheng, M. Claessens and T. Gascoigne et al (eds) (2008) *Communicating Science in Social Contexts: New Models, New Practices*, Springer Publishing, Dordrecht

Wynne, B. (2006) 'Public engagement as a means of restoring public trust in science – hitting the notes but missing the music?', *Community Genetics*, vol 9, no 3, pp211–220

WWViews (World Wide Views of Global Warming) (2009) www.wwviews.org

4

Crossing Boundaries with Deliberation: But Where Are We Going?

Edward Andersson and Thea Shahrokh[1]

In the past three decades researchers have explored how deliberative processes can increase citizen participation in decision-making. The concern is not simply to reverse the decline in traditional modes of participation such as voting, but to also improve the quality and impact of democratic interactions by conducting them in novel ways and in new social spaces. Large-scale citizen deliberations have evolved to address increasingly complex issues, many of which cross national boundaries (Andersson et al, 2010). This chapter examines the origins, characteristics and future implications of five large-scale and cross-national deliberative events: Meeting of Minds (2006), European Citizens' Panels (2006–2007), Deliberative Poll® on the future of Europe (2007), European Citizens' Consultation I (2007) and II (2009) and WWViews (2009). Taken together, these cases reveal a growing body of cross-national deliberative practices that have the potential to foster truly global debates on some of the most urgent issues facing society today.

Deliberative public engagement processes have developed significantly since the early 1980s (Abelson et al, 2003, p240). Until recently these processes have been used exclusively at local and national levels. A more recent development (since 2006) has been the use of these methods in multiple countries to deal with complex issues that cross national boundaries (predominantly at European level). While this is an emerging area of practice, there are now enough cases to warrant an analysis of the results and provide tentative assessments for the future.

The concept of deliberative engagement combines two distinct aspects: citizen engagement in policy, and deliberation. It is possible to have increased public engagement without deliberation, e.g. a referendum (Matsusaka, 2005); similarly it is possible to have increased deliberation among policy-makers without involving citizens. A number of driving factors over the past decades, including declining turnout at elections (IDEA, 2004) and the complexity of many policy issues, has led to an increased shift towards both deliberation in policy and towards greater citizen involvement in decision-making (Wilsdon and Willis, 2004, p39).

In looking at contemporary examples of cross-border public deliberations, this chapter aims to articulate the purposes that this type of citizens' deliberation serves and understand where it is heading. Traditionally, deliberative democracy was only possible on a small scale, through micro-deliberations most commonly addressing community-based issues. The conventional practice for getting 'public input' for larger issues continues to be opinion polls, despite the widely known limitations of that approach (Brodie et al, 2009).

The shift towards cross-national deliberations aspires to a significant evolution of democracy. A brief historical analysis enables us to understand how deliberative methods have been used in a cross-border context, and to contextualize the five international deliberations that we will analyse in this chapter. In order to frame this analysis we will also briefly present the rationale for deliberation, and analyse the complex, interconnected nature of the challenges facing decision-makers that have prompted these cross-boundary deliberations.

A central aim in this chapter is developing categories for analysing cross-national consultations so that this emerging practice can be better assessed and supported.

There are differences among these cases as well as important common themes. For example, participants tend to be selected using random or quota sampling, which ensures that the process reflects a very large target population. In all cases emphasis is placed on the provision of balanced and accurate information to participants and creating a deliberative space to discuss and understand the arguments and come up with their own views on the issues presented. Also, in all cases face-to-face engagement is the key mechanism for deliberation, in some cases with online engagement as a supporting aspect. While the WWViews process took place on a truly global scale, all the other processes have occurred on a European scale.

The chapter concludes with a discussion of trends that may emerge in the next five to ten years. Based on the existing processes, we hypothesize that future deliberations are likely to be both larger in scope and more decentralized in structure, tied closer to concrete policy issues and decision-making structures, and more effective in their use of new communication and social media technologies.

Background to deliberation

In this section we provide an outline of the concept of deliberation and a brief history of deliberative methods. We also discuss the social and political drivers of deliberation and provide an overview of some deliberative methods that have inspired the development of recent transnational deliberation processes.

History of deliberation and democracy

The idea of deliberative democracy, that public policy decisions can and should be reached through informed discussion among citizens, has been a basic tenet of some forms of democracy since Athens in the age of Pericles (Heierbacher, 2007, p102). Deliberative democracy amounts to an emerging socio-political 'movement' where there is a desire by the public for more productive consideration of political issues. Particularly prevalent in the democracies of Europe and the English-speaking world, this desire has been stoked by a frustration with the perceived inadequacies of competitive democracy (Briand and Hartz-Karp, 2009, p169).[2]

Habermas (1929–) advanced the idea of a public world built upon mutual communication and reason, his concept of 'communicative action' emphasizing the utilization of the way humans think and use language to understand one another and plan for common action. Rawls (1921–2002) revived the idea of basing political thinking and action on moral argument. He supported the practice of providing the public with clearly articulated alternative views and options so that people could couple rigorous thinking with personal values to make wise choices (Heierbacher, 2007, p103).

There is disagreement among theorists of deliberation as to what any one definition should be (Button and Ryfe, 2005). Mendelberg (2002, p153) gives a meaningful working definition adapted from Habermas: that true deliberation is when people rely on reasoning that speaks not just to them but to everyone affected by the situation at hand. Briand and Hartz-Karp (2009, p169) explain that what they refer to as 'public deliberation' is a 'maximally inclusive form of political discourse with a problem solving orientation, a discourse in which citizens collectively – even cooperatively – analyse a "problem"'. Deliberation provides an opportunity to cultivate a reciprocal, rational, practical and unprejudiced exchange among individuals. The result of this is a more empathetic view of the other – all others; a much broader understanding of one's own interests; and a better-informed perspective on public problems. In contrast, more competitive democratic models have focused on self-interest in negotiation, and as a result have fallen short of serving the public good (Mendelberg, 2002, p153).

Of course, deliberation is not a panacea. Research on deliberation and deliberative democracy tends to be guided by normative ideals and thus has been criticized as idealistic. The main criticism is that deliberative democracy is not practical in the current large-scale, complex democratic system (Przeworski, 1991). Further criticism that stems from social psychology literature, arguing that deliberation does not necessarily produce more fair outcomes, and even though the aim may be to reach more socially just decisions, deliberative processes are often biased against socially disadvantaged groups such as racial minorities and women. This criticism is usually accompanied by empirical evidence, for example research on small group dynamics has produced findings that deliberation sometimes favours the pre-existing views of the majority (Schkade et al, 2000, p1140) and maintains racial conflicts (Mendelberg and Oleske, 2000, p185).

It is important also that the intentionality behind deliberation with the public is clear. In disentangling the different rationales underlying deliberation it is useful to consider three distinguishable types; normative, instrumental and substantive (Fiorino, 1990, p227). The normative view argues that deliberative processes should occur because they are simply 'the right thing to do'; this means that often there is no reference to the ends in question. The instrumental view aims for public participation to secure particular ends, for example building the reputation of the commissioning organization. The substantive position of public deliberation emphasizes that engagement processes can in fact improve the quality of decision-making processes. Wilsdon and Willis (2004) emphasize the importance of public deliberation being substantive, particularly when issues are intractable and complex. Substantive deliberation is a means to build broader questions and possibilities into the problem-solving process. Greater legitimacy can be established in this situation as decisions are being shaped, not informed (Wilsdon and Willis, 2004, p39). The type of deliberation discussed in this chapter is mainly substantive. Citizen participation in the policy process can contribute to the legitimization of policy development and implementation only if the focus of the outcome is the publicly deliberated evaluative criteria resulting in genuine qualities. Participation, in this respect, can be understood as helping to make democratic structures more sustainable in the long term (Stirling, 2008, p271).

The political dynamics of increased participation

It is important at this juncture to look further into some of the drivers for deliberation, and probe how they are shaping the discourse. We focus below on those that are most relevant to the field of deliberation with the public today, namely: distrust between citizens and the state, creating an increased gap between people and policy-makers; and the increased complexity of issues faced by government. We will consider how each driver has influenced and increased interest in testing deliberative processes.

Increased distrust creates a gap between citizens and the state

Norris (1999, p6) outlines a long-term slide in political trust at the national level, giving examples of how the British have experienced the rise of a sceptical electorate; how Swedish surveys documented a 30-year erosion of trust in politicians; and that widespread cynicism about government embedded itself in Italian and Japanese political cultures, fuelling pressures in the early 1990s for major reforms of the electoral and party systems in both countries. It is prevalent that citizen distrust of elected officials and cynicism towards government is on the increase. There is also the widely held view among policy-makers and economists that citizens hold fast to uninformed opinions and mostly operate from self-interest. A widespread concern is that citizen engagement is fuelling 'NIMBYism' (not in my backyard), an unconstructive tendency to block necessary developments out of pure self-interest, undermining the greater good (Schively,

2007). As a result, the gap between people and the decision-making processes that affect their lives continues to widen.

However, more and more policy-makers are looking for ways to rebuild legitimacy in the institutions of government, with a large proportion focusing on evidence that there is a role for public engagement in policy, service delivery and accountability with regards to bringing citizens and government into a closer relationship. Matt Leighninger (2006) has looked at the traditional relationship between government and the citizen using the parent/child metaphor. In Leighninger's view we are in the mid-transition to an adult–adult relationship characterized by more equal and mature interactions. This is not an easy transition, either for citizens or civil servants who see well-established views challenged on both sides. However, it does represent an opportunity, as we shape and are shaped by this change in relationships, to establish forms of governance that are efficient and egalitarian, deliberative and decisive.

Increasingly complex challenges face society

It is important to also consider that the challenges that governments are dealing with have become more complex and wide-ranging. Societal problems are becoming less bounded and more intractable. States are no longer living independent lives where 'high politics' only dominates international relations.[3] Cooperation is essential on a multitude of transnational issues, whether political, economic or social. Interdependence and the rise of 'non-territorial issues', such as the degradation of the environment, terrorism, migration or communicable diseases, have propelled governments to the centre of new complex challenges that cross national boundaries (Melissen, 2003, p8).

The challenges outlined above are seen by many as shaping the increased involvement of public engagement in decision-making processes. The general public is often left without a clear understanding of how decisions are taken on the international stage. For example, arguably since the process of climate policy began in 1992 there has not been a meaningful voice representing the public directly. On such complex issues there is a strong argument presented by social scientists and several policy-makers to balance the vocal special interests that may dominate decision-making with the involvement of citizens. This argument is founded on the view that by establishing a shared ownership over decisions, a greater sense of shared responsibility may emerge, strengthening governments' ability to tackle similar issues. As Fung (2003, p351) explains, deliberation can contribute to the efficacy of public policy as publics have the opportunity to criticize and consider justifications; this discussion creates an opportunity to enhance the legitimacy of a policy and thus citizens may be more likely to comply or cooperate.

Deliberative methods

Proponents of deliberation assert that the process by which the product of deliberation is achieved is integral to the legitimacy of the final decision. Traditional (non-deliberative) methods of engagement such as publicly open meetings of government bodies, where citizens can only comment on specific issues under consideration,[4] or public opinion polls, have limitations as a meaningful way for citizens to engage in governance. These methods make it difficult for decision-makers to determine the complexities of the various viewpoints at the table. Public meetings, for example, are often subject to the manipulations of organized interests or vocal minorities.

There is a broad grouping of deliberative processes that have been developed for use at the local and national levels that serve to overcome these limitations. These include: planning cells,[5] Citizens Juries®[6] and consensus conferences,[7] to name a few (see explanations below).

As Abelson et al (2003, p243) outline, individual methods may differ with respect to specific features such as participant selection, the type of input obtained or the number of meetings; however, what is consistent within all formulas is the deliberative component where participants are provided with balanced information about the issue being deliberated, and encouraged to challenge the information and consider all views before making a final recommendation for action. The 1970s saw the creation of structured deliberative public engagement processes; previously deliberation played important roles in civic life, but there were no named processes which specifically aimed to get citizens to deliberate on an issue in a structured way to influence policy-making. In 1972, the first planning cell was run in Germany, followed two years later with the first Citizens Jury in the US. By the 1980s governments at the national level were taking more interest in these techniques. In 1984 a Citizens Jury was conducted with governmental sponsors in the US for the first time.

In Europe and North America particularly, recent years have seen the growth of public deliberation to encompass more comprehensive citizen deliberations; for example the development of consensus conferences in the 1990s by the Danish Board of Technology (DBT). These citizen deliberations enable a comprehensive look at the policy issue by involving not only citizens, but also experts, the media and policy-makers. They are a means by which the needs of decision-makers and the public can be met, in particular with regards to issues that are complex in nature, and where meaningful, in-depth dialogue is needed to gather the collective wisdom of a constituency of users. Strategically designed, the models have been developed in a way that has enabled other practitioners to replicate and scale up the design so that very large groups can contribute to the collective wisdom of citizens by enabling them to simultaneously participate in intimate, face-to-face deliberation.

Deliberative Polling®[8] attempts to incorporate a deliberative process into the traditional opinion poll, which ultimately allows for a much greater number of participants than the purely face-to-face methods focused on above by increasing the number of people involved to a statistically significant sample (Fishkin and Farrar, 2005; Fishkin, 2009). Developed by James Fishkin in the early 1990s, the Deliberative Poll combines the strengths of a large representative, random sample while providing opportunities for deliberation over a two- to three-day period. Using pre- and post-deliberation polls, empirical research suggests that participant views do change as a result of the deliberative process; however, it is unclear precisely how this occurs (Abelson et al, 2003, p243).

In the late 1990s the organization America*Speaks* spearheaded a new method known as the 21st Century Town Meeting™,[9] which has further increased the maximum possible number of people simultaneously involved in deliberative processes to many thousands (Lukensmeyer and Brigham, 2002). 21st Century Town Meetings are forums that combine small-scale face-to-face deliberations with the impact and power of large-scale interactions and collective decision-making.[10] This methodology is designed to overcome the common trade-off between the quality of discussion and size of group involved through innovative use of technology.

Analysis and methods

Deliberation would seem to offer a number of benefits as a strategy to influence cross-border decisions. In the remainder of this chapter we will focus on examples of cross-national deliberation as they stand today; focusing predominantly on Europe as this has been where such practices have been developed. We will look at the design, profile and outcome of these deliberative processes in order to assess whether there is an evidence base as well as a conceptually viable future for cross-border deliberations.

As we shall see, many of these ambitious models of cross-border deliberation have been successful despite many challenges that are unique or more substantial than those experienced at national or local levels, including those faced by practitioners working in these environments, the complexities of operating simultaneously in very different legal, cultural and institutional contexts and within funding cultures which are very nationally focused. The processes provide a useful backdrop for understanding the value of processes such as the WWViews in future.

Since 2006, a small but growing number of cross-border simultaneous deliberations have built on the models reviewed in the previous section. We have identified five separate projects which are deliberative, involve citizens, cover issues that cross national boundaries and where the process either takes place in multiple countries at once, or involves participants from many different countries coming together in one location. These five are:

- Meeting of Minds (2006);
- European Citizens Panels (2006–2007);
- Tomorrow's Europe (2007);
- European Citizens' Consultation I (2007) and II (2009);
- World Wide Views on Global Warming (2009).

Our basic questions in analysing these deliberations are: What are the central characteristics of cross-national deliberation and how have these changed over time? To address these questions, we draw on the secondary data and available literature (for example project reports and evaluations) on our cases. The following characteristics are examined in order to understand the common and unique features of each case:

* rationale and purpose;
* funder;
* delivery bodies – organizations who delivered the process;
* participants – who and how many;
* countries;
* methodology – including use of technology; and
* results – product, recipients, expectations, reception.

A meta-analysis of the five cases is presented in the next section, focused on their developmental drivers. This analysis is further informed by interviews with five project managers who were involved in some, but not all, of the projects. We will complete the chapter by discussing what guidance the meta-analysis has given us on directions we should pursue in future to further this field.

Meta-analysis – comparing the approaches

Table 4.1 provides a brief overview of the five processes that we have presented. Detailed descriptions of the cases are presented in the Appendix at the end of this book. The following discussion will analyse how the cross-border processes compare, with a particular focus on how the preceding European deliberations measure against the most

Table 4.1 *The development of cross-boundary large-scale deliberation, 2006–2009*

Deliberation	Topics	Number of:		
		Participants	**Countries**	**Events**
Meeting of Minds (2006)	Brain science	126	9	30
European Citizens' Panel (2007)	Rural issues	337	9	9
Tomorrow's Europe (2007)	Social and foreign policy issues for Europe	362	27	1
European Citizens' Consultations (ECC), 2007 and 2009	Energy and environment, family and social welfare, immigration; EU economic and social future in a globalized world	*1500 *figures per deliberation*	*27	*28
WWViews (2009)	Climate change	4000	38	44

recent global deliberation: the WWViews process. We will also highlight possibilities for the future in the field of cross-national deliberation.

Similarities

Looking at the processes outlined above we have identified a number of similar cross-cutting themes and principles that seem to define current practice in cross-national deliberation. This is important as it helps to establish an understanding of good practice in this area and will allow us to forecast any trends in the development of this type of process.

All cross-border deliberations mentioned have had a dual purpose
They have all discussed actual policy topics ranging from rural issues to brain science. Alongside the discussion of policy the projects have all also had the aim of piloting a methodology and developing good practice in deliberation. Because deliberation at the international level is still a novel field, there is much innovation and methodological development. It is likely that as cross-border deliberations become more prevalent, the issues discussed will become more important and the method used less so. The fact that the European Citizens' Consultation (ECC) was funded again in 2009 perhaps points to a more sustained and institutionalized approach in the future.

All cases were based on deliberative processes previously used at the national level
Most projects were clear about the source of their inspiration, be it Citizens Juries, 21st Century Town Meetings or Deliberative Polling; the aim was to replicate deliberative quality. In most cases the methodology was a hybrid, adapting and combining elements of different methods to create a tailored approach. In the case of WWViews the process was largely based on DBT's work in the field of consensus conferences.

In all cases except Tomorrow's Europe there were elements of national or regional interaction preceding international interaction
Several processes emphasized the importance of participants being able to get together with people from the same cultural background and who speak the same language first before engaging in multilingual environments. For WWViews all events took place at the national or regional level with no cross-border face-to-face events.

All processes have involved participants selected using some form of random selection with demographic sampling criteria to ensure a mixed group of people who would not normally have taken part in discussions on the topic
Given the size of the populations affected by cross-border issues it is not surprising that organizers have reverted to sampling as opposed to open access or stakeholder

processes. This methodological aspect is shared across all projects we have looked at and aims to reach beyond a self-selecting group and bring new perspectives to issues.

All processes also involved small-group discussions as the main form of participant interaction

This is in keeping with practice in most deliberative methods emphasizing the importance of face-to-face interaction. The best practice principles that Abelson et al (2003) mention for local and national deliberations hold true at the international level too.

All processes involved complex coalitions and funding arrangements

Unlike national deliberations, which are often funded by one or two sources and which have one clear lead organization, the coordination of international projects require a number of diverse partners to work together collaboratively. The WWViews process represents a significant increase in the number and diversity of partners involved. Some countries involved had significant experience with deliberative processes in decision-making while others exemplified authoritarian decision-making. Sourcing funding for the process in such highly variant policy cultures creates a challenge (Rask et al, 2009, p282).

Differences

However, there are also important differences between the processes. Reviewing the differences is crucial to understanding the emerging trends for future deliberation; which areas need improvement and how successes can be concretized in future.

The various projects had very different levels of impact and policy influence

In many ways the European level has a clearer set of institutions to influence compared to the institutional setting for COP15. In the European cross-boundary cases the projects were commissioned by parts of the European Commission with responsibility for the topic area. Some, such as Tomorrow's Europe, had strong links to influential policy-makers and decision-makers whereas others have had less clear access points. Overall, the processes tended to have very limited direct influence over policies. Transnational decision-making tends to be more complex and opaque than national processes and this may have contributed to the limited impact.

The WWViews process differs from all the other cross-border processes in that it did not involve interaction among participants from different countries

The European processes involved a central event where all or a selection of participants intermingled, involving simultaneous interpretation. For WWViews, interaction between national events was not planned centrally but was at the discretion of each partner. In light of the argument that deliberations mixing different nationalities pose complications (e.g. the organizers of Meeting of Minds state that multilingual dialogue takes up a lot of time and requires specific interpreting and facilitation skills – see Steyaert and Vandensande, 2007, p43), the WWViews approach to interaction among different national deliberations offers an alternative that warrants consideration.

Cultural differences between participating countries were much larger at the WWViews event

Notwithstanding the diversity of EU countries, they are all functioning democracies that share many common rules through the European Union and its treaties. The countries that took part in WWViews were far more diverse.

Trends

These pioneering projects have sown the seeds for more sustained international deliberations in the years to come. Notwithstanding many differences among them, key areas of convergence are evident (Levine et al, 2005). Future processes are likely to continue to involve centrally set information, involve randomly selected participants, and meeting at the national or regional level at face-to-face events. There also seems to be a direction of progression for cross-border deliberations, which points to how future processes might differ from what has been done to date.

Bigger and more sophisticated cross-border deliberations

Meeting of Minds involved nine countries and scarcely more than 100 citizens. A few years later WWViews involved 38 countries and nearly 4000 participants. This is a remarkable increase in scale in a short period of time. It is unlikely that all cross-border processes will involve this many countries, but at the very least it has been demonstrated that it is possible to involve thousands of people on multiple continents. In light of this success and the increasingly global character of social and environmental problems it is likely that even larger numbers will be seen in the near future.

A move towards using communication and social networking technology

Processes like WWViews and ECC in 2009 have included wider online interactions before, during and after events in order to allow wider input to and engagement with the issues. Information technologies have radically increased the number of people that can be involved simultaneously. Numerous pioneering e-engagement processes show us that it is possible to engage meaningfully with large numbers online. Increasingly cross-national deliberations will include both high-quality controlled deliberations and mass online engagement. Interestingly, present trends in funding within the European Commission point to the increased funding of online deliberative projects; the Pan European eParticipation Network (PEP-NET), for example, is among recent measures that will increase online activity within 'the future' of cross-border deliberation.[11] Utilizing new technologies to link up dispersed local-level events still faces a number of challenges, but can in the future add much value to cross-border or large national engagement processes.

Increasing connectivity to policy-making

Most cross-national deliberations have focused on general principles and visions with no direct link to decision-making. In order to make a substantive impact on live policy issues, future international deliberations are likely to be more closely tied to decision-making institutions. For example the second ECC process in 2009 had a much more targeted topic and allowed for clearer policy proposals to emerge than the first ECC process in 2007. It was perhaps inevitable that early processes would be quite visionary and divorced from decision-making, since they were pilots. The WWViews process was the first cross-border deliberation to focus solely on influencing one specific summit or event. Proponents of cross-national deliberation seem to have grown confident in the methodology involved, and as such have the capacity to move the design and focus of the process more concretely towards policy impact, although the complexity of transnational decision-making may make this more difficult.

In addition, linked to the trend above, an opportunity opens up for this kind of deliberative process to impact on multiple levels of government simultaneously. Most prominently the ECC deliberations tried to do this by involving national politicians in national events and EU politicians in the final EU event. In a slightly different way, WWViews also looked for this multilevel impact by encouraging national organizers to lobby their country's COP15 delegates for commitments to climate change targets; the aim being that this would impact on global decision-making in the long run. As mentioned, it is often more effective to tackle these kinds of complex problems at multiple levels of government. Cross-national deliberation appears to be embracing this path.

Decentralized structures, distributed funding

The high costs of deliberative processes may lead to their being largely decentralized to networks, supported by discussion kits, a central resource and an online feedback mechanism. In the case of WWViews the process as a whole was not funded centrally but each country had to find their own money; this distributed funding model is likely to be more common in the future as the processes get larger and more complicated.

Institutional consolidation

The majority of cross-national deliberations to date have been one-off pilots. This is perhaps unsurprising as they have been grand experiments in deliberation. After at least six successful processes we can clearly state that international deliberations are workable. In the future it is likely that at least some multilateral organizations or decision-making structures will start to look at cross-national deliberations as part of a structured institutional response. While governments and multilateral organizations have the necessary funding to pay for these processes and the decision-making power to give them force, the involvement of charitable foundations and non-governmental organizations (NGOs) in designing, running and funding deliberations tends to lend itself to more credibility from the public. It can be generalized that governmental institutions have more clout while NGOs and foundations provide more public trust.

The examples discussed show how we are seeing a growing practice of cross-border citizen deliberation in response to complex decisions about issues of international significance. With WWViews what was a purely European affair has become truly global. This phenomenon has occurred as a result of governments and other national and international institutions needing to manage the fact that common challenges are transnational but public opinions remain nationally based. WWViews was the first citizen deliberation on a truly global scale (Bedsted and Klüver, 2009, p3). It is in many ways unique, but as this chapter shows it builds on and improves the practice developed in cross-border deliberations at the European level. Many of the organizations involved in WWViews had previously been involved in these European projects, for example DBT, which was a partner in the Meeting of Minds process. The past five years have been a period of astounding methodological development for cross-border deliberation. Future practice will be much improved from looking at good practice from multiple sources, locally, nationally and internationally.

Conclusions

In an increasingly interconnected and interdependent world, citizen influence over complex societal problems at the national level can be enhanced by global deliberations that address issues transcending the authority of individual nation states. The cases examined in this chapter suggest that cross-national deliberative processes can play an increasingly significant role in good global governance, while also highlighting challenges that will have to be addressed in order to live up to that potential. The WWViews process represents the most impressive cross-national deliberation to date in terms of its scope, ambition and complexity; but it is unlikely to be the final word in deliberative evolution. In the coming years there are signs that new global deliberations will seek to harness the power of online communication and social networks; build closer ties to decision-making institutions; and develop organizational models that effectively configure decentralization and centralization. If there is progress in meeting these objectives, these deliberations will further propel the field of practice towards a truly global debate.

Notes

1   Both authors were part of the team that organized the UK WWViews Event: Edward Andersson as overall project manager and Thea Shahrokh with responsibility for event logistics.
2   Competitive democracy can be described as political actors trying to win contests with others who oppose their efforts. 'Moreover, political actors do not shy away from making use of

resources ... that generate political influence or power to obtain from the political process outcomes that they or their clients want, irrespective of the substantive merits of their decisions' (Briand and Hartz-Karp, 2009, p169).

3 'High politics' outlines the understanding that national security imperatives drive state behaviour.

4 These 'public meetings' are not designed with a commitment to take on board participant feedback and feed it into policy-making. As such they are often seen to limit two-way interaction between decision-makers and the public as well as deliberation among participants. They can also self-select participants who are particularly confident, articulate and motivated about particular issues, thus ensuring their attendance.

5 An explanation of the purposes and methods of a planning cell is available at http://www. cipast.org/cipast.php?section=1018

6 Registered trade mark symbol used only once at this point, but should be considered as in use throughout the chapter. Visit the Jefferson Centre for more information on the Citizen Jury; www.jefferson-center.org.

7 To explore the consensus conference further, please see www.tekno.dk/subpage.php3?article =468&toppic=kategori12&language=uk.

8 Registered trade mark symbol used only once at this point, but should be considered as in use throughout the chapter. For more information on Fishkin's Deliberative Polling visit: http:// cdd.stanford.edu/polls.

9 Unregistered trade mark symbol used only once at this point, but should be considered as in use throughout the chapter. For further information on the 21st Century Town Meeting see http://www.americaspeaks.org.

10 The methodology of the 21st Century Town Meeting combines small-group discussion (10–12 people) with large-scale collective decision-making through the use of networked laptops. Ideas generated by the public are then coded by a team of professionals to develop priorities. Participants then vote on these priorities, enabling a close connection between the larger group and the discussions within the small groups. Information on this method was taken from www.peopleandparticipation.net.

11 For more information please visit http://pep-net.eu/blog/tag/inenglish.

References

Abelson, J., Forest, P.-G., Eyles, J., Smith, P., Martin, E. and Gauvin, F.-P. (2003) 'Deliberations about deliberative methods: Issues in the design and evaluation of public participation processes', *Social Science and Medicine*, vol 57, no 2, pp239–251

Andersson, E., Burall, S. and Fennell, E. (2010) *Talking for a Change*, Involve, London

Bedsted, B. and Klüver, L. (2009) *World Wide Views on Global Warming, From the World's Citizens to Climate Policy-Makers*, Danish Board of Technology

Briand, M. and Hartz-Karp, J. (2009) 'Institutionalizing deliberative democracy: Theoretical and practical challenges', *Australasian Parliamentary Review*, vol 24, no 1, pp167–198

Brodie, E., Cowling, E. and Nissen, N. (2009) *Understanding Participation: A Literature Review*, Pathways through Participation, http://pathwaysthroughparticipation.org.uk/wp-content/ uploads/2009/09/Pathways-literature-review-final-version.pdf

Button, M. and Ryfe, D. M. (2005) 'What can we learn from the practice of deliberative democracy?', in J. Gastil and P. Levine (eds) *The Deliberative Democracy Handbook: Strategies for Effective Civic Engagement in the 21st Century*, Jossey-Bass, San Francisco, CA

Fiorino, D. (1990) 'Citizen participation and environmental risk: A survey of institutional mechanisms', *Science, Technology & Human Values*, vol 15, no 2, pp226–243

Fishkin, J. (2009) *When the People Speak: Deliberative Democracy and Public Consultation*, Oxford University Press, New York

Fishkin, J. and Farrar, C. (2005) 'Deliberative polling: From experiment to community resource', in J. Gastil and P. Levine (eds) *The Deliberative Democracy Handbook: Strategies for Effective Civic Engagement in the 21st Century*, Jossey-Bass, San Francisco, CA

Fung, A. (2003) 'Survey article: Recipes for public spheres', *The Journal of Political Philosophy*, vol 11, no 3, pp338–367

Heierbacher, S. (2007) 'Dialogue and deliberation', in P. Holman, T. Devane and S. Cady (eds) *The Change Handbook: The Definitive Resource on Today's Best Methods for Engaging Whole Systems*, Berrett-Koehler Publishers, San Francisco, CA

IDEA (2004) *Voter Turnout in Western Europe: Since 1945*, International Institute for Democracy and Electoral Assistance (IDEA), Stockholm

Leighninger, M. (2006) *The Next Form of Democracy: How Expert Rule Is Giving Way to Shared Governance – And Why Politics Will Never Be the Same*, Vanderbilt University Press, Nashville, TN

Levine, P., Fung, A. and Gastil, J.(2005) 'Future directions for public deliberation', in J. Gastil and P. Levine (eds) *The Deliberative Democracy Handbook: Strategies for Effective Civic Engagement in the 21st Century*, Jossey-Bass, San Francisco, CA

Lukensmeyer, C. J. and Brigham, S. (2002) 'Taking democracy to scale: Creating a town hall meeting for the twenty-first century', *National Civic Review*, vol 91, no 4, pp351–360

Matsusaka, J. G. (2005) 'Direct democracy and fiscal gridlock: Have voter initiatives paralyzed the California budget?', *State Politics and Policy Quarterly*, vol 5, no 3, pp248–264

Melissen, J. (2003) *Summit Diplomacy Coming of Age*, Discussion Papers in Diplomacy, Netherlands Institute of International Relations 'Clingendael', www.clingendael.nl/publications/2003/20030500_cli_paper_dip_issue86.pdf

Mendelberg, T. (2002) 'The deliberative citizen: Theory and evidence', in M. X. D. Carpini, L. Huddy, and R. Shapiro (eds) *Political Decision-Making, Deliberation and Participation*, Series, No 6 Research in Micropolitics, JAI Press, Greenwich, CT

Mendelberg, T. and Oleske, J. (2000) 'Race and public deliberation', *Political Communication*, vol 17, no 2, pp169–191

Norris, P. (1999) *Critical Citizens: Global Support for Democratic Government*, published by Oxford Scholarship Online, 2003

Przeworski, A. (1991) *Democracy and the Market: Political and Economic Reforms in Eastern Europe and Latin America*, Cambridge University Press, New York

Rask, M., Lammi, M., Repo, P. and Timonen, P. (2009) *Empowering Tomorrow's Consumers Through WWViews*, Future of the Consumer Society, Finland

Schively, C. (2007) 'Understanding the NIMBY and LULU phenomena: Reassessing our knowledge base and informing future research', *Journal of Planning Literature*, vol 21, no 3, pp255–266

Schkade, D., Sunstein, R. and Kahneman, D. (2000) 'Deliberating about dollars: The severity shift', *Columbia Law Review*, vol 100, no 4, pp1139–1175

Steyaert, S. and Vandensande, T. (eds) (2007) *Participatory Methods Toolkit: A Practitioner's manual. European Citizens' Deliberation Methods Description*, Meeting of Minds Europe, Brussels, www.meetingmindseurope.org/europe_default_site.aspx?SGREF=14&CREF=6688

Stirling, A. (2008) '"Opening up" and "closing down": Power, participation, and pluralism in the social appraisal of technology', *Science, Technology & Human Values,* vol 33, no 2, pp 262–294

Part III
Evaluating the WWViews Process and Results

5

Deliberating or Voting?
Results of the Process Evaluation
of the German WWViews

Rüdiger Goldschmidt, Leonhard Hennen,
Martin Knapp, Christiane Quendt,
Nadine Brachatzek and Ortwin Renn

Governmental, non-governmental as well as private organizations increasingly pay attention to participatory and deliberative methods for involving the public in decision-making processes. WWViews is most generally characterized as the first-ever global citizen consultation, building on the growing number and variety of deliberative events at local, regional and national levels (Joss and Bellucci, 2002). In addition to the political dimension of having citizens of 38 countries voice their informed preferences about policy actions towards combating climate change, the WWViews process constituted a unique opportunity for conducting research and gaining more scientific insights into the structure and procedures of participatory processes. Empirical research in this field still lags behind theory and practice (Carpini et al, 2004, p315).

This chapter sums up some of the empirical findings and conclusions based on the research activities of the WWViews partners in Germany.[1] A major focus lies on the recruitment process and especially the composition of the German WWViews participant group. The question of participant selection is central to the evaluation of the external fairness and the political legitimacy of participatory processes (Webler, 1995, p62; Goldschmidt et al, 2008; summary in Renn, 2008, p282), and has important implications for the design of global consultations that will be addressed in our concluding section.

After a short overview of our research methods, subsequent sections of this chapter address the recruiting process; the composition of the participant sample in terms of socio-demographic variables, as well as participants' attitudes to the subject discussed. The analysis also focuses on core perceptions of the process by the participants, the session observers and the interviewed group moderators. These findings and their implications are further discussed in the last section of the chapter.

Overview of instruments and methods of the evaluation

The analysis is based on a fully standardized survey among participating citizens of the German WWViews and 12 half-standardized telephone interviews with participants conducted two months after the WWViews consultation. This is complemented by insights gained from a focus group discussion with 8 group facilitators (out of 13) conducted three weeks after the event. In addition, a team of researchers observed the event by using a qualitative prestructured observation sheet,[2] and the observations of the organizing team were also incorporated into the analysis.

With respect to the evaluation surveys,[3] two data sets are analysed in this chapter:

• The *main survey* was conducted immediately after the WWViews event and mainly provides information on perceived process performance and the quality of results. Of 81 participants at the German event, 79 completed this survey.
• The *pre-survey* was administered to 88 accepted participants in August 2009, approximately one month before the WWViews event. Respondents to the pre-survey who did not complete the main survey were excluded from the data analysis to ensure the validity of comparisons between the two surveys.[4] This resulted in a total sample of 68 persons.[5]

The results of the surveys, interviews and observations form an expansive body of material, only parts of which are included in this chapter.

The recruitment process

In line with the suggestions of the project coordinator, DBT, there were three options for recruiting citizens for the German WWViews event:

• Request a certain number of addresses from communal and regional civil registration offices (advantage: complete set of data; disadvantage: very time-intensive due to decentralized registration).
• Buy addresses from a commercial marketing company (advantage: fast and easy; disadvantage: very expensive and leads to bias, because only special groups of the population are selected).
• Randomly select addresses from national or regional telephone directories (advantage: less time-consuming than registration offices, cheap, in-house; disadvantage: certain groups under-represented).

In light of its funding and time constraints, the German project team chose the last option.

Another decision that had to be made in advance was the region from which the participants were to be recruited. A national sample would make WWViews a genuinely German citizen conference, but Germany is a rather large country and travel time and expenses can be considerable. Due to funding restrictions, participants for the citizen conference were ultimately recruited from the federal state of Baden-Württemberg in south-west Germany. This federal state is one of the largest in Germany, with about ten million inhabitants. It is characterized both by urban and rural areas, and has regions rich in both technology and agriculture. Baden-Württemberg's large geographical area includes a variety of different landscapes (that might be influenced differently by climate change and climate adaptation measures): e.g. a low mountain range (Black Forest) in the south and a flood plain (Rhine plain) in the middle of the region.

Based on its previous experiences with citizen panels, the project team expected that the return rate on the letter of invitation would be between 2 and 5 per cent. Integrating this data with DBT guidelines, invitation letters were sent to more than 5000 randomly chosen citizens in the region. The organizing team hoped for at least 250 positive answers, from which 120 participants were to be chosen according to criteria such as age, gender, occupation, etc.

However, shortly after the deadline for responding to the letter of invitation, only 70 people had applied. Approximately 4000 more invitations were therefore issued, yielding a total applicant pool of 135 citizens. This left little space for selecting people according to socio-demographic characteristics and, in light of uncertainties over the number of confirmed participants who would actually attend the deliberation, the project team decided to accept everyone who wanted to participate. By the day before the event 30 people had dropped out, mainly due to illness, and ultimately 81 participants attended the event. Since this did not meet the standard of the project coordinators, we will focus our analysis on how the composition of participants may have influenced the process.

While the number of people who dropped out before the event was within the scope of expectations, the low response rate of less than 1 per cent was worrisome. Invitation letters and the recruitment strategy were tested beforehand – they did not contain any language or phrases that were expected to discourage participation. What might have been the reasons for this result? A likely cause is the day of the week WWViews was scheduled. Saturday is typically a 'family, shopping and household day'. Younger citizens were under-represented (see data in next section), and it is likely that many people in this demographic segment who might otherwise have participated were deterred by the full-day commitment. Furthermore, although the project team recruited people from a confined area, travelling time for some of the invited citizens might have been another obstacle to participation, even though all transport costs were reimbursed. Another reason can be seen in the selection mode itself: younger people especially use mobile phones intensively, and these are not usually listed in phone directories. Lastly, the WWViews event took place the day before the German national elections, perhaps making that particular Saturday more important to prospective participants for private rather than civic activities.

Composition of the participant sample

This section reports on the socio-demographic composition of the participants, along with data on their attitudes and preferences. Another focus is on how the project team handled issues arising from the composition of participants in order to ensure the best possible results.

Socio-demographic variables, political engagement and social networks

Age and Gender: According to the organizers' records, the youngest participant was 16, the oldest 89, and the average age of the participants was 50 (N=80). There was an under-representation of participants under 30 (6.1 per cent) and an over-representation of people over 50 (50.6 per cent). The relative proportions of women and men were 40.7 per cent and 59.3 per cent, respectively. The organizers made sure that at least two or three women were placed at each discussion table, and participants were asked to be mindful of this gender disparity in the course of their discussions.

Educational background: According to the survey results, 62.7 per cent of the sample graduated from German secondary school and qualified for university admission ('Abitur'). The share completing the vocationally oriented 'Realschule'[6] and 'Hauptschule'[7] curricula accounted for 13.4 per cent and 11.9 per cent, respectively. Well-educated citizens were clearly over-represented.

The survey results for vocational education confirmed the latter finding since 35.4 per cent of participants had a university degree and 16.9 per cent a degree from a university of applied sciences ('Fachhochschule'), far higher numbers than for the overall German population.[8] About 3 per cent of the sample had gained a PhD degree. In contrast, 24.6 per cent had graduated from a professional school ('Fachschule') or university of cooperative education ('Fachakademie' or 'Berufsakademie') or worked as supervisors in business and industry. Over 12 per cent had completed an apprenticeship ('Abgeschlossene Lehre') and 6.2 per cent had been to technical school ('Berufsfachschule'). While setting up the groups for discussion, the project team managed to have at least two people without higher education at each table.

Employment status: 50.7 per cent of the participants (N=67) were employed full time and 10.4 per cent worked part time.[9] By their share of 17.9 per cent, 'retired people' constituted a third, broadly represented group. Another 4.5 per cent of the surveyed participants were homemakers, and 3 per cent were unemployed, while 6 per cent selected 'other' as their occupational category. The remaining categories were all under 3 per cent.

Migration: 7.5 per cent of the participants were born outside Germany, and 13.4 per cent had at least one parent who was born outside Germany.[10]

Societal engagement and involvement in special societal networks (Figure 5.1): The vast majority of the sample reported regular participation in national elections, but only 8 per cent reported active involvement in a political party, whereas nearly 90 per cent of the participants had no party affiliations. More than 20 per cent of the sample stated that they were actively involved in organizations focusing on political or societal issues. One third of the participants had engaged in societal and charitable activities.

Approximately 60 per cent of the participants reported that they adapt their lifestyle to protect the environment. Based on the organizers' report, 10 per cent were members of 'green organizations',[11] while 6 per cent of the survey sample indicated active involvement with an organization focusing on environmental protection and 8 per cent indicated moderate activity.

Despite the fact that many participants reported personally engaging in environmental protection efforts, most were not formal members of institutions such as parties, interest groups or green organizations. It is important to note that the WWViews event in Germany was not overrun by environmental activists. Nonetheless, the organizers ensured that those involved in green organizations were spread equally over all tables and advised the group facilitators to ensure that green and non-green opinions were involved in the discussions.

Nearly 40 per cent of the participants indicated that they were religious and 20 per cent of the sample had been actively involved in a religious organization. Approximately 46 per cent of the citizens claimed to have sufficient experience to participate effectively in public participation about societal decisions. Overall, the recruitment succeeded in

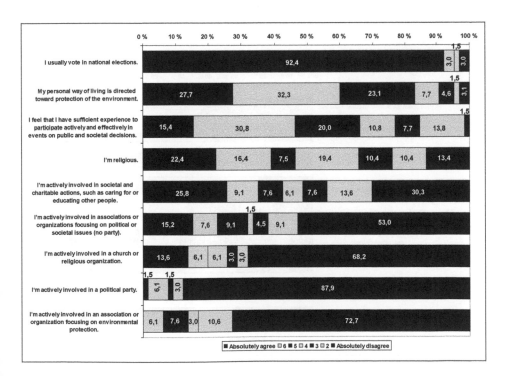

Figure 5.1 *Participant political engagement and social networks, N>64[12] (pre-survey)*

bringing together citizens from a broad variety of social and demographic backgrounds. For instance, the recruitment process achieved a wide range between younger and older participants. The proportion of employed people was rather high. However, there was an over-representation of people over 50, men and educated citizens.

Participants' attitudes and preferences

Motives for participation (Figure 5.2): Most of the participants had clear reasons for attending the WWViews event. A large majority was motivated to enhance their own knowledge about the subject of climate change (learning as main driver). Two sub-categories of learning were cited as motivation by roughly equal proportions of the sample: (i) to learn how to act in a more environmentally friendly manner (pragmatic) and (ii) to enhance one's own understanding of different perspectives (discursive).[13] In addition, a strong driver for the participants was to get more involved in climate change decision-making processes. Overall, the subject of climate change was of central interest to the participants (see next section).

Interest in the subject of climate change (Figure 5.3): Climate change is a relevant subject in the everyday life of nearly half the participating citizens. Nearly half the respondents stated that they like to read books and magazines about this subject. This high proportion is not surprising when looking at the educational levels of the

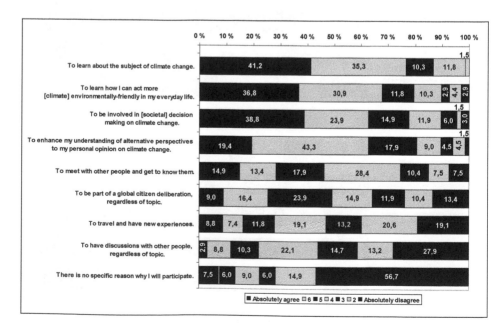

Figure 5.2 *Participants' motivation to attend the WWViews event, N>66 (pre-survey)*

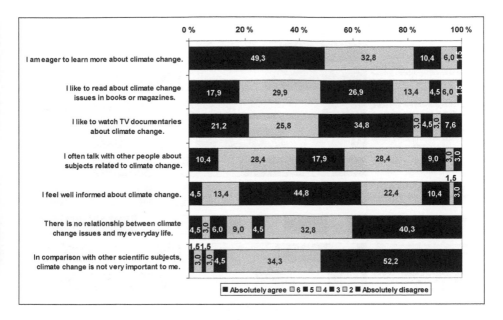

Figure 5.3 *Participants' interest in the topic of climate change, N>65 (pre-survey)*

participants and their interest in learning. Virtually the same proportion of participants liked to watch television documentaries about climate change. Nearly 40 per cent of the participants often talk with other people about climate change. However, only a small majority felt well-informed about the topic, in spite of the fact that the overwhelming majority of the participants emphasized the importance of the subject climate change for everyday life and within sciences.

Attitudes to climate change (Figure 5.4): A clear majority of the participants assumed that climate change will have a negative impact on the quality of life of future generations, although a negative impact on today's daily life was only perceived by a quarter of the sample. The vast majority of participants were aware of the seriousness of the climate change problem. They were convinced that this problem has not been exaggerated by environmentalists. It is not surprising therefore that half the participants supported the claim that society needs to reduce its income and comfort in order to avoid the negative impacts of climate change. Somewhat surprising is the pessimistic view that the participants had of future prospects for global climate solutions.

Only a quarter of the sample perceived Germany as a world leader in reducing energy consumption and mitigating climate change effects. New technologies were seen as an effective instrument for combating climate change by 20 per cent of the participants, whereas less than 15 per cent regarded the impact of citizens as a crucial driver. Approximately 10 per cent of the sample expected a significant positive impact through the advancement of international agreements.

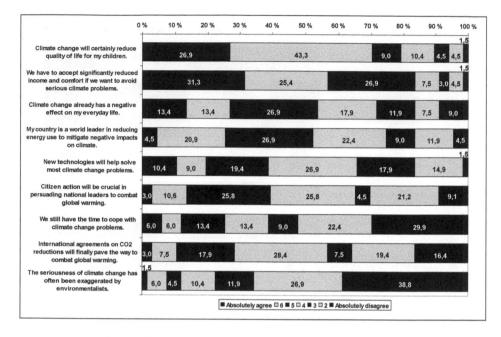

Figure 5.4 *Attitudes to climate change, N>65 (pre-survey)*

Assessing the dialogue process of the WWViews event in Germany

The next section presents assessments of the event process based on the responses of the participants to several performance indicators in the main survey and based on interviews and observations.

The evaluation survey

The participants assessed the event very positively in their responses to the main survey, as reflected in the overview statistics in Figure 5.5. The German event was obviously perceived as a success in terms of efficiency, logistics, productive use of time and organizational aspects such as technical support. In addition, the performance of the moderators was highly appreciated and participants expressed high satisfaction in terms of procedural fairness, which is an essential indicator for the quality of dialogue (cf Webler, 1995, p62; see summary in Renn, 2008, p282; Goldschmidt and Renn, 2006; Blackstock et al, 2007, p734; Goldschmidt et al, 2008). Most participants were also satisfied with the time scheduled for recreation.

Most participants expressed satisfaction with the quality of contributions of other participants as well as the comprehensibility and effectiveness of the information brochure. However, these indicators of competence received less support than the indicators of organizational aspects and efficiency. With regard to transparency indicators, most participants had a clear idea of how the results of the WWViews event would be used in the political and public arenas.

In contrast to these positive responses, 9 per cent of the participants felt there was inadequate time to discuss the issues, and positive responses on this item are much lower compared to the other indicators. Still, a majority of participants signalled their satisfaction with the time allowed for discussion.

There was more ambiguity when it came to the question of whether all relevant societal groups were represented at the event. The approval rate for the indicator measuring structural fairness is much lower than the rate for the variable measuring procedural fairness, and the number of 'don't know's is higher than on average (N=72). It is possible that participants find this item difficult to assess. One might assume that the citizens also perceived some shortcomings in the sample composition, specifically the bias towards highly educated participants. During the post-event interviews, some participants criticized the under-representation of younger people.

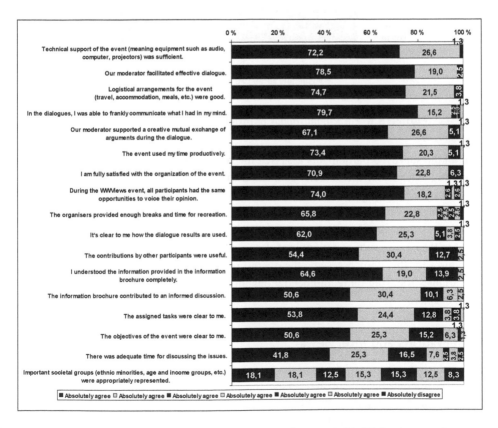

Figure 5.5 *Participant ratings of process quality, mostly N>78 (main survey)*

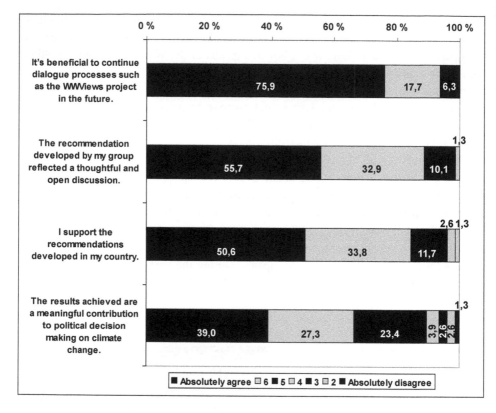

Figure 5.6 *Participant ratings of results quality, N>76 (main survey)*

Overall the surveyed participants assessed the results of the event positively. An overwhelming majority of the citizens confirmed that it is beneficial to continue dialogue processes such as WWViews in future (Figure 5.6). The participants valued the recommendations developed by their own working groups and broadly supported the other recommendations developed at the event. However, they were more hesitant in believing that they had produced meaningful contributions to political decision-making. Responses for this item, assessing the overall results, were less positive than they were for variables focusing on recommendations alone. We explore this matter further later in the chapter.

Observations and interviews

The evaluation survey and the interviews revealed that participants at all times felt they were able to contribute their own perspectives and concerns to the debate. Six insights about the nature and quality of table discussions can be derived from table observations and interviews with participating citizens as well as moderators.

First, the table moderators unanimously reported during their focus group session that, after the initial discussion round, even those participants who were reluctant to articulate themselves became more engaged over time and participated actively in discussions. Most participants needed time to familiarize themselves with the rules and the format of the procedure. It was an advantage that all working phases were structured in a similar manner, resulting in the development of a routine in subsequent rounds. It was also observed that at some tables the participants partially managed to moderate the debate themselves.

The observations confirmed the survey findings that, in general, all participants had opportunities to contribute to the debate, and that most did so actively. However, there were individual differences in knowledge and communicative competence which resulted in more and less influential individuals. More knowledgeable participants assumed the role of 'table experts' who took the lead in inserting facts, dates and concepts when needed.[14] While these knowledgeable participants had a bigger share of speaking time, they did not necessarily dominate the discussion. The dialogue remained fair and open. In other cases knowledge questions were directed towards the moderators, who were expected to know the answers.

Out of 12 interviewees, 9 asked for a stronger involvement of experts to raise the substantive level of discussion at the tables, although they were clear that experts should support rather than lead the discussion. The discussion about emissions goals was specifically mentioned as a place where expert input would be beneficial in guiding participants through complex scenarios.[15]

Moderators reported that participants were content with the presence of controversial statements and did not seek a strong consensus. As observed by moderators, controversial statements and claims 'were just left without comment'.[16] The observation results of the evaluation team confirmed this finding. To some extent, participants did not reflect assessments and contributions even when overwhelming evidence was presented. A possible explanation for this finding is that the event design called for individual votes at the end of all four working sessions prior to the recommendation session and, therefore, established a discursive pattern that focused participants' attention on individual responses to predefined statements instead of on jointly developed group results.

However, controversies among participating citizens did occur when addressing issues such as how to translate demands and expectations into concrete measures, for example, whether CO_2 reduction goals or commitments should be undertaken by developing as well as industrial countries. These controversies did not evolve into productive, informed discourse. The table observations and participant interviews clearly showed that limited background knowledge and the absence of experts impeded a joint exercise to clarify complex issues. Moderators in some cases referred to the information brochure in order to provide additional input, but this was obviously insufficient to meet the participants' information needs. Moderators reported that the participants were tentative and uncertain when asked to decide on the appropriateness of reduction goals, and that they complained about the abstract character of these goals.

A significant issue in the moderator focus group was that the questions answered by participants at the end of each discussion lacked the rich perspectives and statements that emerged in the discussion. The moderators reported that many participants complained about the undue limitation of the standardized response options. In addition, 3 of 12 interviewees expressed their unease with regard to the voting processes at the end of each discussion round. They felt that predefined statements were too abstract and could not cover the complexity and the distinctions of the discussions at the tables. This might explain the survey results presented above rating the recommendations significantly higher than the overall results of the event. Two of the interviewed participants also criticized the recommendations as too general with respect to the measures society should take to combat climate change. This further supports the argument to intensify the dialogue among the participants.

Discussion

Overall, the WWViews event delivered results that were expected by the organizers in a timely, efficient and fair manner. Our concluding remarks will discuss the design of the WWViews event, especially in light of the socio-demographic composition of the WWViews panel, but also the attitudes of the participants. Experience from the WWViews event led to the recommendation that future worldwide citizen participation should be organized with more focus on the model of deliberation rather than decisionistic voting processes.

Reflections on the composition and attitudes of the German WWViews panel

The recruiting process succeeded in involving a broad variety of citizens in the event (although older and highly educated citizens were over-represented), but was less successful in securing responses to the invitation to participate. The low attendance rate of less than 1 per cent of those invited might be explained by the complexity of the topic, thus attracting mostly highly educated people. Another likely cause is that WWViews was scheduled on a Saturday, which might particularly explain the lack of people under 50.

Quite interestingly, the over-representation of educated people among the German WWViews participants was turned to advantage: some of the highly educated participants functioned as quasi-experts based on knowledge from their present or former professional experience, for instance, as an engineer. These quasi-experts brought more novel and distinctive arguments into the discussion than those participants less familiar with the topics.

A high proportion (10 per cent) of participants compared to the German population as a whole were active in social or environmental initiatives or groups.[17] Among the respondents there was a strong awareness of environmental issues and problems: 60 per cent of participants reported that they adapt their lifestyle towards environmental protection. The group composition was characterized by environmentally aware, better-educated individuals of the German population. However, the German WWViews event was not overrun by environmentalists and members of green organizations.

When asked about their motivation to participate, a strong majority of the participants expressed the clear desire to learn more about the subject. Two different sub-motives were revealed: (i) to learn how to act in a more environmentally friendly manner (pragmatic component) and (ii) to enhance their own understanding of different perspectives (discursive). Participants were also interested in being involved in decision-making on climate change (involvement).

The classic finding, that ordinary citizens feel (and are) uninformed about complex scientific topics (Slovic, 1987, p285) seems also true for the WWViews participants. However, nearly half the participants reported watching TV broadcasts or reading books about the subject. The equal approval rates for these two variables may be a result of the high level of education of the participants. Almost 40 per cent of the participants confirmed talking with other people about climate change.

The participants' subjective perception that climate change is relevant to them and their children resonates with a recent national survey (Bundesministerium für Umwelt, Naturschutz und Reaktorsicherheit, 2008). A strong majority of the WWViews participants (70 per cent) believe that climate change will reduce the quality of life of their children, while a clear majority of Germans in the nationwide survey think that climate change will have serious negative consequences.[18] However, only a quarter of the WWViews participants expressed experiencing any negative effects of climate change on their personal life. Most respondents thus perceived climate change as a moderate problem for the present, but a serious threat for the future.

Regarding the ways to respond to the climate change, a finding worth noting is that only 10 per cent of the participants expected that international agreements on CO_2 reduction would play a significant role in combating climate change. In contrast, 20 per cent thought that new technologies could provide solutions, and 14 per cent of the sample perceived citizen action as an effective instrument.

Overall, the subject of climate change and especially the opportunity to learn something about this subject was a core driver for attending the event. Climate change is definitely perceived as a relevant challenge for the future and partly as a subject for everyday discussion. However, the citizens themselves felt broadly uninformed about climate change, although generally well-educated citizens were over-represented among the participants, so this might help explain rather pessimistic and restrained assessments of survey questions on how to solve the problem of climate change.

Reflections on the design of the deliberation process

The event design fostered efficiency and transparency of the overall process by offering a standardized procedure for structuring all working phases. However, the relationship between deliberation and voting in the chosen event design raises important questions. The working sessions were generally organized as deliberative group discussion at tables, but, when divergent views emerged during the working sessions, the table rounds often declined to discuss them. Other than the explicit development of a final recommendation in the final working round of the event, the debates were mostly not continued until some kind of common result was achieved. A major reason can be seen in the event design, which stipulated that each debate should end with a vote, based on individual assessments instead of jointly developed decisions. This mixture of starting with a deliberative perspective and completing the process in a 'decisionistic fashion' (Renn, 2008, p11)[19] prompted some critical observations by the participants and the moderators.

Given this partial dissatisfaction with the process, we would recommend increasing the proportion of deliberation in the design of future processes. This recommendation is based on three arguments.

First, the process was designed to complete each working round with an individual vote based on a standardized questionnaire with predefined options. This resulted in limited opportunities for an interactive exchange of views and did not encourage participants to reach a common understanding. Relevant arguments were left unexplored and others were raised but not resolved.

In comparison to simple voting results, one of the advantages of deliberation (Susskind and Ozawa, 1983, p265; Middendorf and Busch, 1997, p52; Carpini et al, 2004, p336) is to deliver explanations of why a specific decision was made and to communicate background knowledge. By reducing the rich conversations in WWViews to multiple-choice questions, the event took only limited advantage of one of the main benefits of deliberation.

However, an argument against strengthening deliberation can be found in the challenge of how to synthesize recommendations coming from a great number of countries – in the case of WWViews from a total of 38 countries. A solution could be to create a multilevel participation process that combines national consultations with an international convention of spokespeople where final recommendations can be drawn up. Online discourses could support this multilevel process design. Multilevel designs were already tested in participatory exercises at European level. It should be noted that these initiatives were considerably funded by the European Commission. WWViews followed a completely different approach.

Second, increasing the deliberative component of future WWViews events can foster knowledge transfer, which would satisfy the strong desire of the participants to learn more about the subject. An overwhelming majority of the participants perceived a need for information about climate change in the pre-survey, but scientists were not involved as consultants in the table discussions. Rather the organizers used standardized infor-mation that was checked by experts beforehand. This strategy of standardized informa-tion input improved the comparability of results, but left the less-informed participants with an uncertain grasp of the complex issues. Fortunately, there were enough highly educated participants in place to address this need, at least in part. A more deliberative design with experts would offer the opportunity for an interactive 'real-time access' to relevant knowledge claims. This leads us to a basic finding, derived from experience in other projects outside WWViews: lay persons are basically competent to judge, but it's a major challenge to design a process that ensures that lay persons are able to develop informed discussion and informed decisions on a complex problem.

A final argument is that the tension between deliberation (aimed at achieving under-standing and developing an informed, common result) and voting (collecting informed preferences from the participants) places incompatible demands on the recruitment process (Webler, 1995, p61; Middendorf and Busch, 1997, p52). Deliberation demands a sufficient representation of all relevant arguments in the dialogue. Recruitment should therefore ensure adequate variation with respect to the participants' arguments, opin-ions and backgrounds, which was substantially achieved in WWViews Germany. For developing a valid snapshot of preferences, however, the voting procedure requires statistical 'sampling'. This implies a representation of positions in accordance with their overall distribution in society, rather than a complete representation of arguments irrespective of their distribution in society.[20] The recruitment process demonstrated that a representative sample of the German population is almost impossible to accomplish for a participatory exercise. Using random selection, the response rate for invitations to deliberative events normally varies between 1 and 15 per cent. This makes it nearly impossible to recruit a sample that mirrors the actual proportions of the population. So, in the end, there is no meaningful model of how to interpret individual votes in such a participatory process.

For these three reasons, we recommend that future efforts to organize worldwide citizen participation should be inspired by a model of deliberation rather than decision-istic voting processes.

Notes

1 Leonhard Hennen (Institute for Technology Assessment and Systems Analysis (ITAS)), Rüdiger Goldschmidt, Nadine Brachatzek and Ortwin Renn (University of Stuttgart) worked together in the research team. Martin Knapp and Christiane Quendt (ITAS) organized the WWViews event in Germany.
2 Two observers changed their tables after each working session to observe another table. One observer remained at the same table for the entire event (Table A).
3 The survey questionnaire is comprised mainly of full standardized items with seven answer options. Generally, the most positive rating option for each statement is coded '7' and the

most negative '1'. The socio-demographic variables were measured with reference to stand-ard survey instruments for the German population.

4   A third survey wave, the 'post-survey', was in process when this chapter was completed, possibly leaving out of this analysis some participants who may be included in future data analysis.

5   A control analysis based on the full data set of the pre-survey supported the main findings.

6   This German educational achievement is equal to secondary modern school in the UK and high school in the US.

7   This German school teaches basic skills for people who will go into vocational training or the unskilled labour market.

8   According to Federal Statistical Office data (2008), 7 per cent of Germans hold a uni-versity degree and another 5 per cent have a degree from a university of applied sciences (www.destatis.de/jetspeed/portal/cms/Sites/destatis/Internet/DE/Content/Statistiken/BildungForschungKultur/Bildungsstand/Tabellen/Content100/Bildungsabschluss, templateId=renderPrint.psml, accessed 27 February 2010).

9   According to Eurostat, the proportion of employed to unemployed German citizens is 70:7 per cent, using 2008 data (http://epp.eurostat.ec.europa.eu/tgm/table.do?tab=table&init=1&p lugin=1&language =de&pcode=tsiem010, accessed 27 February 2010).

10   Approximately 20 per cent of the German population is comprised of first or second genera-tion immigrants (www.destatis.de/jetspeed/portal/cms/Sites/destatis/Internet/DE/Navigation/Statistiken/Bevoelkerung/MigrationIntegration/MigrationIntegration.psml, accessed 27 February 2010).

11   According to the process documentation, 8 of the 81 participants (9.9 per cent) were members of the main German environmental protection organizations (BUND, NaBu and Greenpeace). A few participants might be members of local or smaller groups.

12   All diagrams provide data for one set of variables on a related theme; the variables are ranked from those with the highest to the lowest proportion of responses in the two most positive response options ('Absolutely agree' and '6'). Values below 1 per cent are not displayed in the diagrams.

13   The two most positive responses per variable totalled 67.7 per cent for the pragmatic motiva-tion, and 62.7 per cent for the discursive motivation.

14   Two engineers participated in the working group of Table A. The engineers reported from their former projects and brought a lot of background knowledge into the dialogue. They also supported the group when clarifying terms and specifications. However, an older electrician (lower level of education) contributed actively, drawing primarily on his life and work experience.

15   All 12 interviewed citizens reported that the WWViews information brochure was very helpful as a balanced and non-biased source that they had used in preparing for the event (see also results of the main survey). However, responses from the moderator discussion as well as impressions from the observers reveal that this information actually played a marginal role in the discussions. Reference as an input for discussion was made predominantly by modera-tors, and many participants admitted to not having read through the material at all.

16   The quoted comment was made by a moderator in the focus group after the event.

17   According to a survey based on the general population of the federal state of Baden-Württemberg (N=1886), in 2004 4 per cent of the population were volunteer workers in the

sector 'protecting the environment/nature/animals' (Hoch et al, 2007, p47). Although there is a conceptual difference between active members of organizations and volunteer workers, this result indicates that the proportion of WWViews participants active in the environmental protection sector is higher than in the average population.

18 There are some methodological differences between the WWViews evaluation surveys and the cited population-wide survey. The cited survey offered only four response options, which is significantly different from the seven response options of the evaluation survey analysed in this chapter. However, in the cited survey only 4 per cent strongly and 18 per cent moderately agreed with the statement: 'There are no serious negative consequences of climate change', with the vast majority responding that there are negative consequences (Bundesministerium für Umwelt, Naturschutz und Reaktorsicherheit, 2008, p25).

19 The decisionistic model is one of three types for collective decision-making. The other two are technocratic and participatory. The decisionistic model assumes that factual information about potential consequences of action is provided by scientists, while elected politicians have the duty to evaluate these consequences according to their understanding of public desirability or in line with collective preferences. It is more inclusive than the technocratic model because it builds on a division of labour between experts and politicians rather than deferring to the experts. The decisionistic model is less inclusive than the participatory model, however, as it limits the number of relevant actors to two, whereas participatory approaches also include civil society groups. The term 'decisionistic' in the discussion above is used to describe the choice of predefined response options by the (expert) organizers instead of a development of these options by the WWViews participants themselves.

20 The operational target of the recruiting process should therefore be greater than 100 participants, but this increases costs and logistical concerns that are especially salient in global deliberations, where the funding resources of participating countries vary widely.

References

Blackstock, K. L., Kelly, G. J. and Horsey, B. L. (2007) 'Developing and applying a framework to evaluate participatory research for sustainability', *Ecological Economics*, vol 60, no 4, pp726–742

Bundesministerium für Umwelt, Naturschutz und Reaktorsicherheit (Federal Ministry for the Environment, Nature Conservation and Nuclear Safety) (2008) *Umweltbewusstsein in Deutschland 2008: Ergebnisse einer repräsentativen Bevölkerungsumfrage*, Silber Druck, Niestetal, www.umweltdaten.de/publikationen/fpdf-l/3678.pdf

Carpini, M. X. D., Cook, F. L. and Jacobs, L. R. (2004) 'Public deliberation, discursive participation and citizen engagement: A review of the empirical literature', *Annual Review of Political Science*, vol 7, no 1, pp315–344

Goldschmidt, R. and Renn, O. (2006) *Meeting of Minds: European Citizens' Deliberation on Brain Sciences, Final Report of the External Evaluation* (Stuttgarter Beiträge zur Risiko und Nachhaltigkeitsforschung), Universität Stuttgart, Stuttgart, www.kbs-frb.be/uploadedFiles/KBS-FRB/Files/Verslag/ECD_Finalreport_ExtEval_complete.pdf

Goldschmidt, R., Renn, O. and Köppel, S. (2008) *European Citizens' Consultation Project: Final Evaluation Report*, (Stuttgarter Beiträge zur Risiko und Nachhaltigkeitsforschung),

Universität Stuttgart, Stuttgart, http://elib.uni-stuttgart.de/opus/volltexte/2008/3489/pdf/ AB008_Goldschmidt_Renn_Koeppel.pdf

Hoch, H., Klie, T. and Wegner, M. (2007) 'Lebendige Bürgergesellschaft in Baden Württemberg: Ergebnisse des zweiten Freiwilligensurvey', *Statistisches Monatsheft Baden-Württemberg*, no 2, pp44–49

Joss, S. and Bellucci, S. (eds) (2002) *Participatory Technology Assessment: European Perspectives*, University of Westminster, London

Middendorf, G. and Busch, L. (1997) 'Inquiry for the public good: Democratic participation in agricultural research', *Agriculture and Human Values*, vol 14, no 1, pp45–57

Renn, O. (2008) *Risk Governance: Coping With Uncertainty in a Complex World*, Earthscan, London

Slovic, P. (1987) 'Perception of risk', *Science*, vol 236, no 4799, pp280–285

Susskind, L. and Ozawa, C. (1983) 'Mediated negotiation in the public sector: Mediator accountability and the public interest problem', *American Behavioral Scientist*, vol 27, no 2, pp255–279

Webler, T. (1995) 'Right discourse in citizen participation: An evaluative yardstick', in O. Renn, T. Webler and P. Wiedemann (eds) *Fairness and Competence in Citizen Participation: Evaluating New Models for Environmental Discourse*, Kluwer Academic Publishers, Dordrecht

6

Reflecting a Citizen Consultation Project from a Technology Assessment Perspective: The Case of Austria

Ulrike Bechtold, Michael Ornetzeder
and Mahshid Sotoudeh

Most climate researchers currently argue that global warming is caused by human activities (Oreskes, 2005); climate policy, however, is confronted with conflicting knowledge and values. WWViews provided a unique opportunity to explore citizens' perceptions of global warming in different cultures and contexts and to incorporate a broader base of values into global climate policy. Although a full resolution of the many conflicts of climate change may not be feasible even through a highly participatory approach, decisions can be better legitimized if they are negotiated more transparently and supported by an open dialogue process. WWViews aimed to do this by incorporating a group that has not had a direct say in this political process so far, global citizens; not to act as a replacement for the established decision-making system, but to complement it.

WWViews Austria was organized and conducted by the Institute of Technology Assessment (ITA). As a department of the Austrian Academy of Sciences in Vienna, ITA has traditionally emphasized academic research. The Institute has many years' experience of stakeholder participation (e.g. Delphi Austria in 1998; see Aichholzer, 2002), and the role of participatory technology assessment (pTA) has increased in recent years (e.g. projects focused on Austrian technology policy, the role of end users in sustainable energy technology and the changing connections between privacy and public security; see Nentwich et al, 2006, 2008; Bechtold and Sotoudeh, 2008; Cas, 2010). While many of ITA's pTA projects have been pan-European, WWViews expanded these activities to the global context.

In this chapter we analyse and discuss our experiences with WWViews Austria as the organizers of the national citizen consultation. We first review different conceptions and characteristics of pTA and then focus on some of the key issues contributing to the different levels of success and legitimacy of the Austrian WWViews process. In conclusion, we draw some practical lessons for future activities like WWViews.

According to Joss and Bellucci (2002, p5), the term participatory technology assessment (pTA) refers to a 'class of methods and procedures of assessing socio-technological issues that actively involve various kinds of social actors as assessors and discussants'. Among European practitioners, pTA is generally understood as a means of involving different actors with the purpose of improving the credibility of decision-making processes on both factual and evaluative levels (Bellucci et al, 2002, p273). These definitions apply to WWViews: its focus was on the assessment of COP15 negotiations; it involved socio-technological issues such as environmental and economic impacts of resource consumption, emissions reduction and technology transfer; and its goal was to improve the COP15 decision-making process by providing an additional source of insights on a matter that concerns the public.[1]

Another perspective on pTA emphasizes the function of citizen participation to provide a forum of communication and social learning for different actors (Renn et al, 1995). This perspective emphasizes the role of 'pTA as a facilitator to "reconfigure the interface between decision-maker and affected" (mode of inclusion) next to its capacity of delegating trust to those who participate (mode of delegation) and revealing different views (mode of mediation)' (Hansen, 2006, p575). From this perspective, and in order to support effective dialogues between different actors, it is important to pay attention to the social context of the pTA process.

In our analysis of the Austrian WWViews we will answer the following questions, each of which adopts a different perspective on WWViews as a pTA process:

- What was the societal and political context of the Austrian WWViews?
- What were the goals and functions of WWViews as a pTA exercise?
- How was the Austrian WWViews process legitimized, both internally and externally?
- To what extent did the Austrian WWViews attain procedural neutrality?

In the following four sections we shall reflect on these questions in the light of recent literature on pTA. In the final section we conclude with some practical suggestions for future WWViews-type deliberations.

The societal context of pTA

To practise pTA involves considering a number of contextual aspects. Saretzki (2003) has emphasized that citizen participation depends largely on the socio-political context, which means that political traditions profoundly shape the means of pTA. According to Bellucci et al (2002), the societal context of pTA includes the technology innovation system, the political system in general and the public sphere, including the question of whether there is technology-related controversy at stake. Those preconditions not only influence the impact of pTA results, they also influence the willingness of citizens to participate (Grunwald, 2000). We therefore focus on the national socio-political context of WWViews Austria by analysing international and national addressees and external timelines as contextual factors that proved particularly important for WWViews Austria.

The socio-political context of WWViews Austria

In recent years climate change has become a widely discussed topic in Austria, in part as the result of severe floods in the Danube region in 2002, 2005 and 2006, which many observers ascribed to climate change. By 2007, numerous radio and television stations had launched special programmes focusing on climate change.[2]

As a political issue, climate change gained importance in the political arena some years ago when it became obvious to insiders that the country would fail to meet the targets required by the Kyoto Protocol, to which Austria was a signatory nation. As a consequence, in July 2007, the Austrian climate and energy fund was established, which was the first time that research and technology diffusion was sponsored to a notable extent by the government, with the explicit aim of reducing greenhouse gas emissions on the national level. In spite of this major step and other national and regional measures in recent years, it seems inevitable that Austria will fail to meet the reduction targets (Anderl et al, 2009). In the face of this unfavourable political situation representatives of the ministry responsible felt constrained to support an even stronger position for COP15, as proposed by some countries.

In parallel with the mainstream consensus on global warming, climate change discourses in Austria – as in most other countries – have been marked by significant controversy. For example, in internet forums the existence of climate change is still questioned or dismissed as a conspiracy invented by environmental activists.[3] These opposing discourses, however, have had little impact on public opinion, which continues to reflect considerable concern about global warming (Reiter, 2008).

Regarding the culture of participation in Austria, there is a long tradition of involving major interest groups and stakeholders in political decision-making processes. Even if this neocorporatist form of balancing of interests (the so-called *Sozialpartnerschaft*) has become less important in the last decade, it is still the dominant model for representing various societal interests in policy-making. Forms of direct and participatory democracy have been used in sporadic, and sometimes even manipulative, ways at the national level (Pelinka, 1999). However, on the municipal level we can find striking although rare examples of participatory policy-making (Ornetzeder and Kozeluh, 2004).

Only recently have some organizations started to consistently promote public participation activities. The most prominent example is the work of the Strategic Group on Participation,[4] an independent think tank supported by the Federal Ministry of Agriculture, Forestry, Environment and Water Management (Life Ministry). This group aims to raise awareness of participation among the general public as well as politicians, civil servants and business actors, and it recently published a manual on participation targeted at practitioners.

The primary responsibility for climate issues in Austria is vested in the 'General Environmental Policy' division of the Life Ministry. The head of this division served as leader of the Austrian delegation to COP15. Two offices within the division – 'International environmental issues' and 'Emission control and climate protection' – have been involved in developing national climate policies. Some corporatist organizations (e.g. the industrialists' federation) have also formulated their positions regarding climate change and representatives of these organizations were part of the delegation to COP15. While these players have little experience with citizen participation, their staffs have generally responded favourably to the participatory activities of WWViews Austria.

In pTA, the addressees of a deliberative process constitute an important contextual factor, since the intended audience inevitably shapes the definition of the goals of the process. Moreover, addressees decide how and to what extent the results from pTA exercises will be incorporated into parallel decision-making processes. Precise goal definition and specification of addressees, according to Bütschi and Nentwich (2002), are the two major preconditions for designing both the deliberation process and its external communication.

On a national level, the Austrian delegation to Copenhagen and Austrian climate policy-makers were identified as the addressees for our results. As already mentioned, the administrative units concerned showed some general interest in national WWViews activities. However, the utility of the WWViews outcome was limited for these actors, since the Austrian position for COP15 was already agreed upon within the EU member states (for which reason these actors also declined to fund WWViews).

In order to reach our addressees despite these difficulties, we therefore addressed the Austrian media and sought to increase the project publicity. However, Austrian mainstream media were not interested in reporting on this participatory procedure – even though we highlighted the role of WWViews as the first worldwide pTA project ever. Except for some specialized print media, there seemed to be no significant 'market' for such news. This situation was to some extent compensated for later, in the context of the COP15 conference, when some level of interest in Austrian WWViews was raised within the national mass media.

To sum up, we could not identify a national audience that was expecting the results of WWViews Austria. Furthermore, we had to cope with a dual structure of addressees characterized by a mixture of national and supranational policy-makers and various representatives of the media.

Political timing

Political timing is another important context and success factor for pTA processes. Results of pTA, it is argued, should be available while they are still relevant for political decisions (Abels and Bora, 2004b); at the same time, pTA processes should not take place too early (Skorupinski and Ott, 2002). In this vein, Grunwald has argued that there are optimal 'spaces of societal framing of technology', in which pTA can make significant contributions to policy (Grunwald, 2000).

The Austrian position for COP15 was formed in an EU-wide process and more-or-less fixed long before the WWViews event took place.[5] Therefore, the results of the national WWViews event were not really able to influence this specific political decision.[6] However, since COP15 did not lead to a binding agreement, and because negotiations for a new agreement at the EU and international level continue, WWViews results will remain relevant and can potentially inform and actually influence national climate policy decisions.

Although the COP15 timeline was poorly aligned with WWViews, the Austrian delegates to COP15, with whom we sustained contact, were nonetheless interested in the process. Before 26 September 2009, the Life Ministry provided us with a statement of the Minister on the relevance of WWViews.[7] The Minister (along with key staff) also visited the WWViews deliberation and press event.

Functions of pTA

A number of functions attributed to pTA are connected to the choice of appropriate methodology (Slocum, 2003; OECD, 2004). Abels and Bora (2004a) make a distinction between conflict-resolving and advisory pTA. In our view WWViews fits the category of policy advice and is methodologically bound to the ideals of deliberative and discursive democracy (Dryzek, 2000; Stirling, 2004). In this context, we recognize the provision of new insights, impact on decision-making and social learning as the most salient functions of pTA, which we turn to next.

Provision of new insights

Generation of new insights based on citizens' and societal stakeholders' (as distinct from experts') knowledge and value judgements is an important function of pTA processes (Grunwald, 2004). As Failing et al (2007, p49) emphasize: 'While there is little utility in strictly defining the lines between scientific and local knowledge or traditional and contemporary knowledge, the principles of decision analysis suggest that it is useful to distinguish whether the knowledge claim is fact-based or value-based.'

Since WWViews was focused on generating information on citizens' opinions about various climate policy questions, the project resulted in a unique database of ordinary citizens' values and preferences, both nationally and internationally. A concrete example is the question in the deliberation asking whether fossil fuel prices should be increased in order to limit consumption and therefore also limit greenhouse gas emissions. (In Austria a substantial majority of participants, 61 per cent, agreed with this measure that would ultimately affect their own lives by increasing fuel prices.)[8]

WWViews did not aim to broaden the factual knowledge base of climate change issues. Nevertheless, the citizens' recommendations can be regarded as a source of knowledge that is mediated by their values. The recommendations indicated popular concerns related to climate change and climate policy, as in Austria, where the top three recommendations called for expanded education as a precondition to participation, public support, transfer of CO_2-efficient technology and an effective global climate agreement adjusted to diverse national capabilities (Box 6.1).

Box 6.1 The three top-ranked recommendations of WWViews Austria

EDUCATION AS PRECONDITION FOR PARTICIPATION IN DECISIONS

1 The true victims of climate change must be integrated into the decision-making processes.
2 Unrestricted access to education and information is a precondition for effective participation in such decision-making processes.

RESEARCH FOR THE BENEFIT OF ALL!

We recommend increased research on cost-efficient low-carbon technologies and support for global technology transfer. These technologies should not be developed with a view to profits, and must be affordable for everyone.

KYOTO – WHAT FOLLOWS?

Global, mandatory, effective implementation of necessary climate guidelines that use technology transfer and energy and emission standards to solve, rapidly and on a long-term basis, the global problem of greenhouse gas emissions – within the limits of the possibilities of each country.

It is crucial to acknowledge that the recommendations add knowledge not previously extant about citizens' views, based on informed deliberation among peers. In effect, therefore, WWViews and pTA more generally can contribute to new and relevant insights based on the unique knowledge-making practices of informed deliberation.

Impact on decision-making

The outcomes of a pTA exercise should inform and possibly have an effect on decision-making processes. However, Hennen (2002, p259) reminds us that 'there are serious cognitive, normative and pragmatic restrictions to a direct influence of results of TA processes'. Indirect impacts, in turn, are difficult to trace and hard to evaluate. One strategy to handle this situation is to broaden the view of what the impacts of pTA could be (Bellucci et al, 2002). The outcomes of a pTA exercise, for example, can facilitate societal framing of problems by identifying and characterizing crucial questions (Grunwald, 2004). Another strategy is to increase the impact of pTA by connecting it more closely to the policy process (Abels and Bora, 2004b; Hennen et al, 2004).

The most obvious limits to political impacts of the Austrian WWViews were inherent in the timing of the project and the political system itself. First, the low political impact can be explained in terms of the tendency of Austrian decision-makers to regard participation as a means to transfer information from experts to an uninformed public. Therefore, participation processes are reduced to exercises in environmental education along the lines of the much-criticized 'deficit model' (Sturgis and Allum, 2004) that assumes the citizenry is ignorant and needs to be educated by experts and leaders. The scepticism voiced by administrative persons against participation (Depoe et al, 2004) held true also for the administrative assessment of WWViews Austria. Second, organizational barriers such as the rigid decision-making structures were potential obstacles to a consideration of the results of WWViews within the Austrian climate political process.

Nevertheless, even if the results of WWViews were not directly used in national policy processes, informing the political decision-making process at the national (and international) level about the WWViews goals and activities took many forms. The relevant inputs for the COP15 negotiations were presented to the national delegation in November 2009 (the Danish Embassy in Vienna was involved and actively supported the project). In order to foster the political impact, media work was emphasized from the beginning of the project, including for example a media workshop, press releases, delivery of professional photo materials and so forth. This resulted in four newspaper articles, three broadcast notices as well as 21 articles published online (including ten press release texts published via the Austrian press agency). Although a well-founded background story on the process or its results was lacking, the process still attracted some interest. However, contrary to our expectations and the efforts we made, the media interest remained low.

Social learning

Social learning is a function of pTA that takes place at multiple levels. On the procedural level it refers to the micro-level processes where people interacting with each other contribute to a common societal discourse (Renn et al, 1995, p9). The process goes beyond the individual and results in a process of transformative change by people understanding themselves and others (Webler, 1995). Social learning in pTA processes can also foster the connection between public debates and decision-making processes (Hennen et al, 2004).

Social learning can be regarded as a secondary aim of WWViews Austria. However, discussions ranged widely, participants were confronted with different views and values, and the exit survey results clearly indicate that social learning took place: 82 per cent said that other participants' contributions were valuable to them and 77 per cent claimed that their capacity to better understand opinions that differed from their own increased by participating in WWViews. Moreover, 78 per cent of the participants in the exit survey said that they intend to be more climate friendly in their everyday life, and 57 per cent confirmed having done so in a questionnaire administered five months after WWViews. Respectively, 83 per cent expressed their intention to engage more closely in the societal debate on climate change immediately after the WWViews project, and 59 per cent confirmed having done so five months later. Still another indication of participant learning is that the share of people (54 per cent) who said that they were well informed in the pre-survey increased to 88 per cent in the exit survey.

WWViews type of activity can also raise public awareness and fuel broader societal debates on climate change and global climate policy. Considering that in addition to the extensive dissemination activities, contacts were made with policy-makers (e.g. an interview with the Danish Climate Minister) and the internet and social networking media were used to disseminate the event (e.g. national and international project home pages, blogs, photos and video reports on Facebook and YouTube), WWViews Austria was a valuable contribution to social discourse on climate change.

Internal and external legitimacy of pTA

Democratic legitimacy is one of the most discussed issues in the context of pTA. Equitable participation or balanced representation of diverse social actors (Abels and Bora, 2004b; Hansen, 2006) are often mentioned as necessary conditions of successful pTA (Hennen et al, 2004; Rowe and Frewer, 2004). The intention to include 'all the existing normative preferences in the process' as far as possible (Grunwald, 2000, p164) is at the heart of such efforts. Nevertheless, a dilemma of participation may be detected here, because 'exclusiveness as a condition for success collides unavoidably with inclusiveness as a condition for solving the problem of legitimization' (Grunwald, 2004, p120). Hansen (2006) discusses this dilemma in terms of a 'mode of delegation', proposing that in order to gain legitimacy, pTAs must ensure that non-participants perceive participants as legitimate and trusted representatives.

According to Grunwald (2000, 2004) there are different procedural aspects that contribute to the internal (participants' acceptance) and external legitimacy (non-participants' acceptance) of pTA processes. Procedural aspects contributing to internal legitimacy include transparency of who takes part, what are the preconditions to take part in the process and how the actual discussions and deliberations are organized, documented and used thereafter (Grunwald, 2000, 2004; Saretzki, 2003). A successful transfer of 'internal legitimacy' towards the non-participating public is seen as an important condition to 'external legitimacy' by Grunwald (2000, p163). Ensuring that the goals, results and justification of pTA exercises are clearly communicated, are additional factors contributing to both internal and external legitimacy (Saretzki, 2003, p49; Hennen et al, 2004).[9]

Ensuring internal legitimacy

The structure of participants and the preconditions for the WWViews Austria event were clearly communicated to the participants before and during the process. The participants were recruited by selected quota sampling (Schnell et al, 2005). A specialized social research institute conducted the recruiting process and a large number of interviewers all over the country informed potential participants in short face-to-face interviews.[10] There were quotas on age, sex, place of residence, education and current occupation selected to correspond to their representation in the Austrian population. The recruiting process was very successful, since 96 lay people from all over Austria attended the conference and the structure of the panel was quite balanced despite a slight over-representation of persons with higher levels of education. The vast majority of the panellists also had the impression that the most important sections of the population were present (98 per cent).[11]

The participants were not urged to study beforehand (no expertise was required, just personal interest). This was emphasized when potential participants were contacted for the first time as well as later in the invitation letter. Equal access to information was ensured as the brochure was sent to each participant two weeks ahead of the event by mail. The brochure was predominantly seen as balanced in terms of its content. Nearly all participants (98 per cent) said they understood the information material presented in the brochure and 89 per cent estimated that the information contained important insights into the issues of consultation.

A preparatory workshop was designed to ensure the impartiality of the staff (including the process facilitators and other supportive partners) and the procedural and methodological know-how of the team, consisting also of external staff. Since all participants in the exit survey said that they were able to express their opinion easily and 98 per cent of the participants were satisfied with the moderation process, the process in these respects was considered highly successful.

The deliberation process was organized in a way that reflected the commonly accepted rules of WWViews Austria (that were not contested on any occasion by the citizens). The goals of WWViews Austria were clear to 94 per cent of the participants as well as the purpose for which the results were produced (96 per cent). Two types of outcomes were produced: voting results and recommendations containing arguments in a condensed form (the arguments of the participants also played an important role within the discussions ahead of voting, e.g. to clarify the question at stake). Two vote counters tabulated the results in front of the participants, while two reporters documented and entered them in the WWViews web portal. The results were immediately presented to all participants between the discussion slots. A professional translator helped to translate the German recommendations into English and a tailor-made electronic entry mask was used to prevent any losses.

Ensuring external legitimacy

Making the methodology transparent, and hence the way the results are produced, helps transfer the internal legitimacy of pTA to the wider public. Being externally legitimate means that if someone who did not take part in WWViews examined the process critically, they would be convinced that the results were generated in a way that the non-participant (1) would agree with, (2) was sure that (s)he could have been chosen as well and trusted in the mandate of the actual participants and (3) might therefore find the achieved results acceptable.

Measures supporting transparency for non-participants included the publishing of the information materials (each press release contained all relevant URLs) and an explanation of the whole process on the web page.[12] Additionally, information published by the media was supposed to foster transparency for wider publics.

Procedural neutrality

Three important issues to prevent biases and contributing to the procedural neutrality of WWViews are discussed in the following sections: (1) the balance between experts and lay people, (2) goals and their clarity and (3) the neutrality of facilitating the pTA process.

In practice the WWViews process comprised different elements, each of which encouraged participants to express their opinions in different ways. During the first phase of the process the discussions aimed at revealing and discussing different viewpoints (with the help of the table facilitators), but the voting at the end of each discussion round was individual and supported a majority selection process. The formulation of the recommendations in the last phase of the process was a consensus-oriented process, although personal rankings again involved a majority vote.

Balance between experts and lay people

Building a balanced participation procedure requires a consideration of the ways in which lay persons communicate with experts and policy-makers. Neither of the latter groups should dominate in a citizen participation process (Abels and Bora, 2004b; Hennen et al, 2004).

The relationship between experts (or expert knowledge) and lay persons was – as intended by the process design – balanced in a specific way. The process itself and the voting were completely in the hands of the citizens; the whole preparation of the process including the precise formulation of questions, on the other hand, was done predominantly by experts. The only deviation from the intended process was that there was one information provider present at the Austrian WWViews event who answered some questions concerning reduction targets and acceptable temperature increases, and could in this regard be perceived as an expert. The media also asked for this expert's interpretations, since the results alone did not seem to be sufficient for them.

In addition to the 96 citizens in WWViews Austria, there were three visitors who gave their welcome addresses to the citizens: a representative of the Austrian Academy of Sciences, the Austrian Minister for Environment and a well-known climate expert. The addresses took place between the first and the second round of discussions. The visitors were asked to keep the speeches very short and to only recognize the citizens' work and contribution to the deliberation. Unfortunately, two out of three speakers delivered emotional speeches in which they emphasized the citizens' responsibility to change something. Therefore, it is reasonable to assume that the welcome addresses influenced the subsequent voting in which 99 per cent of the participants said that a climate deal is urgent and should be made at COP15. Moreover, all participants from Austria agreed that if a new climate deal was made at COP15, the Austrian politicians should give high priority to joining it. For both questions the average of all countries participating in WWViews was 91 per cent in agreement. Nevertheless, some of the facilitators had the impression that the speeches fulfilled the goal of acknowledging the participants' contributions and boosting morale.

Goals and their clarity

Goals and their clarity according to Saretzki (2003) are of utmost importance in determining the selection of pTA as an appropriate approach. Grunwald (2003) emphasizes that quality control is only feasible in terms of procedural aspects and not when it comes to the validity of arguments. Consequently, it has to be clarified whether the mode of discussion focuses on reaching consensus regarding certain questions or whether the aim is to collect as many different arguments as possible. Voting is another challenging issue. Skorupinski and Ott (2002, p117) argue against it, saying that, 'neither in the factual dimension, nor in the dimension of norms and values is it allowed to end pTA discussions by voting'. In our view standardized questions and voting can be used to generate clear and comparable results at an international level. Voting was organized for WWViews in four thematic areas with 12 specific questions that were directly relevant to the negotiations at COP15. The voting process was designed to support communication of the results to the political decision-makers and the media. This resulted in percentages, which could easily be visualized and introduced as a clear message to policy-makers.

Skorupinski and Ott (2002, p119) suggest three steps for managing pTA discourses: (1) establishing the scope of arguments, whereby the participants shall be allowed to contribute any issue they consider relevant, (2) creatively outlining scenarios and (3) critically evaluating arguments.

In WWViews the first part of the discussion was designed to help participants share their ideas and arguments. In Austria, however, the feedback was that there was not enough time to discuss arguments deeply as suggested by Skorupinski and Ott (2002, p119), even though the recommendations session at the end of the day provided a better opportunity for such discussions. The products of this final session, however, were broad messages to policy-makers without much focus on the supporting rationales.

Neutrality of information and facilitation

Institutions that conduct pTAs should seek neutrality and impartiality to conduct methodologically sound processes (Bellucci et al, 2002; Skorupinski and Ott, 2002). Skorupinski and Ott (2002) argue that maintaining impartiality with information requires that the information material used within the process has to be balanced in terms of its content (pros and cons should be represented) and that equal access to the materials is guaranteed to all participants. In Austria, special attention was paid to ensuring that all participants had an equal access to all information materials.[13]

The neutrality requirement is also directed towards the facilitators who support the deliberation processes. In Austria, a professional and independent facilitator was recruited to support the practical realization of the WWViews process. ITA's role as an impartial academic research institute was another factor that contributed to the neutrality of the citizen consultation process.

Conclusions

The overall goals of WWViews Austria as precisely defined in advance were (1) to contribute relevant input to the COP15 negotiations and national climate delegations (and hence to WWViews as a worldwide project), (2) to hand the results over to the respective decision-makers (political impact) and (3) to publish the results and the process in the media (awareness raising). In our view, these three major goals were achieved to different extents, both on the national and the supranational levels. While direct political impacts were, quite clearly, more limited, WWViews notably contributed to social learning on climate change issues and the global political process of COP15.

As the main aim of WWViews was to inform the negotiating parties at COP15, the project design was focused on the supranational context of these negotiations. Based on our analysis of selected contextual factors of WWViews Austria, however, it seems necessary to pay more attention to the differences between national and supranational (or other national) contexts. An ongoing debate in Austria on climate change supported our activities (e.g. the recruiting process), whereas the external political timing set by COP15 turned out to be a limiting factor for both establishing timely connections to political addressees as well as securing funding for the project.[14]

An important lesson learned from this experience is that pTA arrangements with a complex structure like WWViews have to ensure that major goals of the overall process can be and are adapted to national contexts. In our case, the impact could have been increased by reframing the focus from policy advice towards contributing to national political agenda setting, or addressing a broader range of audiences and focusing on raising public awareness of climate policy. An argument for the salience of the latter functions is that WWViews Austria seemed to be most successful in enhancing social learning in its various forms and in broadening the value basis of policy-making.

In the Austrian case it is clear that the DBT's participation model was transferred to a different policy context without loss in quality. The successful transfer of procedural aspects through a sound methodology is connected to the ability to ensure the legitimacy of pTA. This was reached through a set of procedural solutions: a well-balanced sample was achieved; facilitators enabled fair deliberations; all previously decided preconditions were communicated to the participants and so forth. Assuring internal legitimacy by securing high procedural standards is likely to support external legitimacy of WWViews type of deliberation activities.

Finally, as WWViews was conducted worldwide by many institutions with diverse backgrounds, a closer look at the relationship between the core competence of the organizers and the main goal they choose to reach seems very important. Depending on the capacity of the project manager and the institutional landscape in each country, the goals may be in successful media work, fostering policy pathways for the results, raising awareness, or social learning, to mention only the most obvious. The strict scientific focus of our institute limited our capacities to reach the different targets, and a communications expert would have been an invaluable actor in addition to the scientists involved. We may therefore conclude that it is of decisive importance to reflect on existing competences, find synergies between participating institutions, and support the identification of tailored success criteria. Precise procedural standards – as demonstrated by WWViews – can do much to provide institutional impartiality, but locally based reflections and adjustments are needed to reach a truly successful model of global deliberation.

Notes

1 The controversial nature of global warming is another factor that called for conducting a pTA on this topic (see e.g. Nentwich et al, 2006).

2 Austrian NGOs had warned much earlier that climate change was a human-made problem that threatened civilization, but these warnings were largely ignored by the mainstream media.

3 Critical postings can be found on many websites. Examples are http://blog.viciente.at, www. prasada.at and http://zeitwort.at.

4 The Strategic Group on Participation consists of 25 qualified experts on democracy and participation with backgrounds in many different fields. For more information see: www. partizipation.at.

5 Other countries had similar experiences, and these were discussed at WWViews COP15 events at the Klimaforum (www.youtube.com/watch?v=yaqT1nrzdzQ&feature=related) and at the Bella Center (www.youtube.com/watch?v=FI6oJe_zpQo).

6 The Austrian delegates to COP15, with whom we were in contact, informed us in September 2008 that results generated in September 2009 would arrive too late to impact policy, since the

so-called national positions (which merge into a trans-national EU position) were determined in a series of meetings and conferences long before COP15.

7   The statement from the Minister of Agriculture, Forestry, Environment and Water Management in the exact wording (translation by M.O.): 'Climate protection is the responsibility of politics, economy, industry and every individual person. The negotiations on the decision of the post-Kyoto Protocol in December 2009 in Copenhagen will determine the path to global climate change and define an important new framework. But it is already clear that for implementing all mitigation measures a broad awareness among the public of the need for CO_2 savings, resource conservation and energy efficiency in our personal and professional actions is essential. I therefore welcome initiatives such as the project World Wide Views on Global Warming, which are important multipliers and thus help to achieve this goal'.

8   29 per cent voted for an increase of fossil fuel prices for all countries, 29 per cent only for Annex 1 countries and countries with substantial economic income and/or high emissions, 3 per cent only for Annex 1 countries, 33 per cent voted against a regulation of prices and 6 per cent didn't know or didn't wish to answer.

9   A basic condition for all the foregoing components of legitimacy is that pTA is an element but never a replacement of representative institutions in a democracy (Bellucci et al, 2002; Abels and Bora, 2004a; Hennen et al, 2004).

10 If someone was interested, he or she was asked to sign a mandatory declaration. This process started – because of the summer holidays – 12 weeks ahead of the event. In a second phase the ITA contacted all future participants personally via telephone and started to send out more detailed information material by mail and email.

11 Free will and interest may be regarded as the ultimate motivation to participate, although a stipend of 50 euro added a material incentive.

12 The information brochure and press releases were requested 3594 times via the national WWViews home page (September 2009–May 2010).

13 Some of the participating citizens said that the information videos were biased, and a slight overemphasis on negative aspects was mentioned in the exit survey.

14 Eventually WWViews Austria was financed from our own resources.

References

Abels, G. and Bora, A. (2004a) *Demokratische Technikbewertung*, transcript, Verlag für Kommunikation, Kultur und soziale Praxis, Bielefeld

Abels, G. and Bora A. (2004b) 'Die Leistungsfähigkeit partizipativer Verfahren im Überblick', in G. Abels and A. Bora (eds) *Demokratische Technikbewertung*, transcript, Verlag für Kommunikation, Kultur und soziale Praxis, Bielefeld

Aichholzer, G. (2002) 'Delphi Austria: an example of tailoring foresight to the needs of a small country', in *International Practice in Technology Foresight*, UNIDO, Vienna

Anderl, M., Böhmer, S., Gössl, M., Köther, T., Krutzler, T., Lenz, K., Muik, B., Pazdernik, K., Poupa, S., Schachermayer, E., Schodl, B., Sporer, M., Storch, A., Wiesenberger, H., Zechmeister, A. and Zethner, G. (2009) *GHG Projections and Assessment of Policies and Measures in Austria*, Reporting under Decision 280/2004/EC, Reports, Band 0227,

Umweltbundesamt, Vienna, www.umweltbundesamt.at/publikationen/publikationssuche/ publikationsdetail/?&pub_id=1801

Bechtold, U. and Sotoudeh, M. (2008) *Participative Approaches for Technology and Autonomous Living*, Project report, Institute of Technology Assessment, Vienna

Bellucci, S., Bütschi, D., Gloede, F., Hennen, L., Joss, S., Klüver, L., Nentwich, M., Peissl, W., Torgersen, H., van Eijndhoven, J. and van Est, R. (2002) 'Conclusions and Recommendations', in S. Joss and S. Bellucci (eds) *Participatory Technology Assessment: European Perspectives*, Centre for Study of Democracy, University of Westminster, London

Bütschi, D. and Nentwich, M. (2002) 'The role of participatory technology assessment in the policy-making process', in S. Joss and S. Bellucci (eds) *Participatory Technology Assessment: European Perspectives*, Centre for Study of Democracy, University of Westminster, in association with TA Swiss, London

Cas, J. (2010) 'Privacy and security: A brief synopsis of the results of the European TA-Project PRISE', in S. Gutwirth et al (eds) *Data Protection in a Profiled World*, Springer, Dordrecht, Heidelberg, New York, London

Depoe, S. P., Delicath, J. W. and Aepli, M. F. (eds) (2004) *Communication and Public Participation in Environmental Decision-making*, State University of New York Press, Albany, NY

Dryzek, J. (2000) *Deliberative Democracy and Beyond: Liberals, Critics, Contestations*, Oxford University Press, Oxford

Failing, L., Gregory, R. and Harstone, M. (2007) 'Integrating science and local knowledge in environmental decisions: A decision-focused approach', *Ecological Economics*, vol 64, no 1, pp47–60

Grunwald, A. (2000) *Technik für die Gesellschaft von Morgen*, Campus Verlag GmbH, Frankfurt

Grunwald, A. (2003) 'Zur Rolle der Technikfolgenabschätzung für demokratische Technikgestaltung', in K. Mensch and J. C. Schmidt (eds) *Technik und Demokratie: Zwischen Expertokratie, Parlament und Bürgerbeteiligung*, Leske und Budrich, Opladen

Grunwald, A. (2004) 'Participation as a means of enhancing the legitimacy of decisions on technology? A sceptical analysis', *Poiesis & Praxis: International Journal of Technology Assessment and Ethics of Science*, vol 3, no 1–2, pp106–122

Hansen, J. (2006) 'Public participation: Operationalising the public in participatory technology assessment: A framework for comparison applied to three cases', *Science and Public Policy*, vol 33, no 8, pp571–584

Hennen, L. (2002) 'Impacts of participatory technology assessment on its societal environment', in J. Joss and S. Bellucci (eds) *Participatory Technology Assessment: European Perspectives*, Centre for the Study of Democracy, London

Hennen, L., Petermann, T. and Scherz, C. (2004) *Partizipative Verfahren der Technikfolgenabschätzung und parlamentarische Politikberatung*, TAB-Arbeitsbericht No 96 des Büros für Technikfolgen-Abschätzung beim Deutschen Bundestag, Berlin, www.tab-beim-bundestag.de/de/publikationen/berichte/ab096.html

Joss, S. and Bellucci, S. (2002) *Participatory Technology Assessment: European Perspectives*, Centre for Study of Democracy, University of Westminster, London

Nentwich, M., Bogner, A. and Peissl, W. et al (2006) *Techpol 2.0: Awareness – Partizipation – Legitimität. Vorschläge zur partizipativen Gestaltung der österreichischen Technologiepolitik*, Institut für Technikfolgen-Abschätzung, Final Project Report; Institute of Technology Assessment, Vienna, www-97.oeaw.ac.at/cgi-usr/ita/italit.pl?proj=e15&cmd=get&opt=count

Nentwich, M., Bechtold, U. and Ornetzeder, M. (2008) *Future Search & Assessment 'Energie und Endverbraucherinnen'*, Final Project Report No D33, Institute of Technology Assessment, Vienna, http://epub.oeaw.ac.at/ita/ita-projektberichte/d2–2d33.pdf

OECD (2004) *Stakeholder Involvement Techniques: Short Guide and Annotated Bibliography*, Nuclear Energy Agency, Forum for Stakeholder Confidence, Paris, www.oecd-nea.org/rwm/reports/2004/nea5418-stakeholder.pdf

Oreskes, N. (2005) 'The scientific consensus on climate change', *Science*, vol 306, no 5702, pp1686

Ornetzeder, M. and Kozeluh, U. (2004) *Lokale Agenda 21-Prozesse in Österreich: Neue Formen partizipativer Demokratie?* Study on behalf of Jubiläumsfonds der Österreichischen Nationalbank, Vienna, www.rankweil.at/zwischenwasser/documents/2005/zwischenwasser20050628000012.pdf

Pelinka, A. (1999) 'Direkte Demokratie – mehr als nur Illusion, aber kein Rezept', *SWS-Rundschau*, Heft 2/1999, pp109–119, www.demokratiezentrum.org/fileadmin/media/pdf/illusion_pelinka.pdf

Reiter, E. (2008) *Die Einstellung der Österreicher zu Kernenergie, Klimawandel und Genforschung: Auswertung und Kommentierung der Ergebnisse einer Meinungsumfrage*, Internationales Institut für Liberale Politik, Vienna, www.iilp.at/publikationen/reihe_studien/die_einstellung_der__sterreicher_zu_kernenergie__klimawandel_und_genforschung--258.html

Renn, O., Webler, T. and Wiedemann, P. (1995) 'A need for discourse on citizen participation: Objectives and structure of this book', in O. Renn, T. Webler and P. Wiedemann (eds) *Fairness and Competence in Citizen Participation: Technology, Risk and Society*, Kluwer Academic Press, Dordrecht

Rowe, G. and Frewer, L. J. (2004) 'Evaluating public-participation exercises: A research agenda', *Science, Technology & Human Values*, vol 29, no 4, pp512–556

Saretzki, T. (2003) 'Gesellschaftliche Partizipation an Technisierungsprozessen: Möglichkeiten und Grenzen einer Technisierung von unten', in K. Mensch and J. C. Schmidt (eds) *Technik und Demokratie: Zwischen Expertokratie, Parlament und Bürgerbeteiligung*, Leske und Budrich, Opladen

Schnell, R., Hill, P. and Esser, E. (2005) *Methoden der empirischen Sozialforschung*, Verlag Oldenbourg, München

Skorupinski, B. and Ott, K. (2002) 'Technology assessment and ethics', *Poiesis & Praxis: International Journal of Technology Assessment and Ethics of Science*, vol 1, no 2, pp95–122

Slocum, N. (2003) *Participation Methods Toolkit: A Practitioner's Manual*, King Baudouin Foundation, Flemish Institute for Science and Technology Assessment, United Nations University, Brussels, www.kbs-frb.be/uploadedFiles/KBS-FRB/Files/EN/PUB_1540_Toolkit_11_ScenarioBuilding.pdf

Stirling, A. (2004) 'Opening up or closing down: Analysis, participation and power in the social appraisal of technology', in M. Leach, I. Scoones and B. Wynne (eds) *Science and Citizenship: Globalization and the Challenge of Engagement*, Zed, London

Sturgis, P. and Allum, N. (2004) Science in society: Re-evaluating the deficit model of public attitudes, *Public Understanding of Science and Public Policy*, vol 13, no 1, pp55–74

Webler, T. (1995) 'Right discourse and citizen participation: An evaluative yardstick', in O. Renn, T. Webler and P. Wiedemann (eds) *Fairness and Competence in Citizen Participation: Evaluating Models for Environmental Discourse*, Kluwer Academic Press, Dordrecht

7

Consumerism and Citizenship in the Context of Climate Change

Minna Lammi, Petteri Repo and Päivi Timonen

Most if not all individuals face two roles that they need to play in modern society: the roles of the consumer and the citizen. The interconnection between consumerism and citizenship has been accentuated by developments within nation states and between them. The expansion of consumption in the last three centuries has occurred in a period distinguished by the rise of nation states, political mass mobilization, along with new ideologies and expansion of markets (Brewer and Trentmann, 2006). As the expansion of markets is now proceeding on a global scale, the interconnection between consumerism and citizenship is being newly defined.

Globalization has added to the complexity of this interconnection by expanding these discourses beyond the nation-state context, in which the roles of citizen and consumer became central elements in socio-political organization. This is nowhere more evident than in the arena of climate change policy, where it is hard to envision effective responses unless they engage people and their organizations in all their diverse guises and roles. The WWViews deliberations provide compelling evidence that the participants appreciated the need to break out of sterile formulations of the roles of citizen and consumer.

In this chapter we take a close look at the relationship between consumption and citizenship in the context of climate change policy. We start off with a discussion on the interrelations between consumerism and citizenship, and how these were expressed in the WWViews citizen consultations. Then we proceed to describe and analyse our empirical data which consist of recommendations made by the citizens of 38 countries who participated in WWViews. Finally, we discuss our findings and reflect on the role of consumerism and citizenship in debates on climate change.

Consumerism and citizenship in WWViews

Conceptualizing the relationship between consumerism and citizenship is by no means unproblematic. Stereotypically, consumers are thought to act in their self-interest in markets, whereas citizens are thought to consider social responsibility and cultural community. However, this dichotomy has been challenged by a number of consumer pursuits such as involvement in boycotts and other civic activities (Pattie et al, 2003; Barnett et al, 2005; Soper, 2007). It has also been challenged by concepts like consumer-citizens (see Arnould, 2007; Hilton, 2007), citizen consumers (Vidler and Clarke, 2005; Jubas, 2007; Livingstone et al, 2007) and political consumerism (Micheletti, 2003), which attempt to bridge the conceptual gap between consumers and citizens. These developments in terminology reflect shared interests between consumers and citizens as well as the growing variety of methods of political participation (see van Deth, 2010).

Moreover a consumption-based lifestyle is far more deeply entrenched in society than is usually thought; consumerism and people's attachments to material things precede the post-war age of affluence and fixation on economic growth (Trentmann, 2009). Historically, it is not a new idea to see consumers as a set part of the whole nation. For instance, in post-war Finland people were encouraged to see private consumption decisions in the context of nation building (Lammi, 2006). In Imperial Britain in the 19th and early 20th centuries, consumers were ideally seen as active citizens acting patriotically and unselfishly (Trentmann, 2006, 2008). Still today governments with neoliberal aspirations attempt to frame elements of citizenship in consumption terms, particularly issues concerning public services.

Consumption has also become a way to act politically when lacking political rights (Trentmann, 2006; Jubas, 2007). When citizens are systematically disenfranchised within traditional political channels, they often turn to consumption issues in an attempt to empower themselves. As Frank Trentmann puts it (2006), consumption society was never the given counterpart of markets, but there was an uneven and contested evolution of the consumer where the role of agency and political conflicts energized the consumer in a number of societies.

Tensions about what and how to consume have increased as global markets have become more fully integrated, which is a process where consumers have played an active role (Trentmann, 2009). Ethical consumerism is attributed elements of responsibility and also challenges the stereotypical distinction between consumption and citizenship. In addition to mixing self-interest with responsibility, it also mixes modes of operation: should ethical consumers focus on change through the purchases they make (the consumer mode) or should they opt for change through participation in civic organizations (the citizen mode) (Hirschman, 1970)?

Similarly, it has often been contended that citizenship – or responsible civic mindedness – has given way to individual utility-seeking consumerism (e.g. see Putnam, 1995). This critique is an essential reflective feature of modern consumption (Mayer, 1989; Helenius, 1974), but it has been hard to draw definite conclusions on how much this really is the case (Pattie et al, 2003; Keum et al, 2004). One reason for this is that both consumerism and citizenship have evolved and keep evolving.

It has even been suggested that politics should adopt measures from contemporary consumer culture (Keum et al, 2004; Arnould, 2007). Media, participation and affluence empower citizens and consumers alike. Marketing then plays an important part in political campaigns, and conscious consumers aim to change structures of society (Shah et al, 2007). This would imply that the repertoire of politics has indeed become broader (van Deth, 2010).

In a policy context, the relationship between consumerism and citizenship has long traditions. Consumer policy in general and consumer protection in particular have long been an integral part of the operation of modern states. Of course, there are different ways in which consumers and consumer affairs have been addressed in different states. Trumbull (2006) distinguishes three consumer policy regimes focusing on economic partnership, social interest groups and pressure groups.

Indeed, what comes out of these analyses is that consumerism has been assigned many definitions (Mayer, 1989). At one end of the spectrum we see a neoliberal approach that sees consumption as the result of utility-seeking individuals and accentuates consumption as a desirable driving force of society. Towards the other end of the spectrum we see various forms of 'responsible' consumerism that take into account the negative outcomes arising from consumption. These responsible forms of consumerism operate in ways which are closely linked to citizenship, civic activities and consumer activism, criticizing both contemporary consumer culture as well as shortcomings in political and economic systems.

The debate on climate change is an exemplary indication of how consumerism and citizenship are increasingly blending into each other. Climate change is yet another issue that is becoming an integral part of responsible consumerism. Again, responsible consumerism is manifested in a number of intertwined ways which can be described in terms of types of civic activism (Pattie et al, 2003), repertoires for political expression (Barnett et al, 2005) or options to respond (Hirschman, 1970). Political decision-making is likely to have effects on consumption, and consumption may create or solve political problems relating to climate change.

All this would imply that both consumerism and citizenship are essential civic activities to consider when dealing with issues relating to climate change. We argue that both were indeed considered in the WWViews consultations, although the emphasis was on citizenship. In the process of empowering citizens in the debate and negotiations on climate change, WWViews also came to empower consumers and draw attention to issues relating to consumption and everyday practices.

The WWViews consultations addressed consumers and citizens in terms of participation, methods and outcomes. First, individuals were asked to participate in person, representing all their personal interests. Second, the consultation method represented a political approach to solving a problem which had partly been created by free markets. Third, and most importantly from the point of view of this chapter, the participating citizens deliberated outcomes which directly and indirectly affected them and others as citizens and consumers.

WWViews is unique in providing data on issues relating to consumer and citizen interests on a global scale. Indeed, one strength of WWViews was the detailed documentation of the process and citizen views expressed in it. Although the WWViews consultation was specially targeted to give citizens opportunity to take a stance on the political process, we can now systematically look for issues that are of interest both to consumers and citizens. According to the design of WWViews, these issues stem from a citizen perspective and extend to consumer as well as other perspectives.

The next section elaborates on the ways in which responsible consumerism and citizenship were presented in the WWViews consultations. We pay particular attention to what kind of aspects of consumption were described and hypothesize why so many of the recommendations addressed consumer issues.

Description and categorization of the recommendations

The aim of this empirical section is to introduce themes discussed in the recommendations. The recommendations mainly comprise sets of short sentences. Through an analysis of these recommendations, we can distinguish topics which the participants consider to be of particular importance in climate change negotiations.

In the last part of the WWViews consultations participants were asked to come up with their own recommendations for fighting climate change. As opposed to the main part of the WWViews deliberation, where citizens voted on predefined positions after discussing the issues, in the recommendations section participants were able to themselves define the topics that they considered most important for decision-makers to understand. They endeavoured, according their own expectations and potential, to create the best possible lay-person suggestions about how to fight climate change.

The recommendations were discussed and formulated in the same groups of seven to ten participants that had worked together throughout the consultation day. During the recommendation session, individual participants tested their ideas with other group members holding different views and after discussion reached agreement on one recommendation, which the group facilitators then wrote down.

We consider the recommendation data to have been influenced a priori by the consultation design. Therefore, the recommendations are studied as 'social artefacts', i.e. as produced and influenced by the WWViews arrangements. In this respect, they are neither transparent nor comprehensive representations of the participants' independent opinions, but rather contextually and socially formulated expressions. Our research task is then to interpret the recommendations through textual analysis in order to understand how consumerism presents itself in these expressions. We use content analysis as a descriptive methodological technique to interpret the recommendations.

The WWViews consultation process encouraged the participants to establish stances on mainly COP15 issues, but the process also generated a great number of recommendations on other climate change issues as well. There were altogether 488 recommendations, varying from 3 to 20 by country. Recommendations numbering 15 or more were developed in Australia, Austria, Brazil, Chile, China-Taipei, Denmark, Finland, Indonesia-Jakarta, Indonesia-Makassar, Japan, The Netherlands, Spain, the UK and

US-Arizona. Five or fewer recommendations emerged from Bangladesh, China, France and Russia.

We begin the content analysis by establishing two categories that we call COP15 and non-COP15 recommendations. The COP15 recommendations were directed at issues addressed in the COP15 negotiation process. By contrast, the non-COP15 recommendations were related to climate change but not considered in the COP15 negotiations. Each of the 488 recommendations fits well within these broad yet distinctive categories. The recommendations in the non-COP15 category are of particular interest, because they address issues outside the frame established by the WWViews process, and thus are more reflective of freely formulated stances on consumerism and citizenship than the COP15 recommendations.

The COP15 category included recommendations relating to the topics discussed in the four thematic sessions of the consultations, on which the participants had already answered questions before they began to develop the recommendations. The themes in these sessions were climate change and its consequences; long-term goals, urgency and commitment; greenhouse gas emissions; and the economy of technology and adaptation. These thematic entities were used to code the COP15 category. For instance, recommendations such as 'we support the establishment of a global body which pioneers cultural change to influence and mentor developing nations to lower carbon emissions through an indexed global financial system' or 'It's urgent! Everyone has to take an active part now' were coded as COP15 recommendations in our analysis. Of the recommendations, 282 were coded to the COP15 group, leaving 206 for the non-COP15 group.

The essential interpretive task with the non-COP15 recommendations is to understand how each recommendation describes what politicians and climate policy negotiators should do. We pay close attention to how these non-COP15 recommendations are constructed as a distinctive set of texts. Therefore, we treat these recommendations as qualitative data, noting their meanings and considering how they are expressed in terms of language (Silverman, 2001).

We situate the recommendations in the WWViews context, which we are familiar with as we participated in the organization of the WWViews event in Finland. We read the recommendations as a resource that can be mobilized for many kinds of processes, not just COP15.

The non-COP15 recommendations give us a valuable understanding of the ideas that the citizens consulted have about problems relating to climate negotiations as well as of the limits of citizenship and politics in nation states. When the participants developed their recommendations, they knew these would be published online and read by other citizens.

We analysed the non-COP15 recommendations by coding different ideas related to consumerism and citizenship. The term 'consumerism' is used here to refer to the association between modern consumer society and industrial mass production. According to the tenets of the consumerism and citizenship discussion, we used codes such as 'activism', 'affluence', 'economy', 'knowledge', 'lifestyle', 'media', 'participation' and 'technology' to identify potential themes and concepts. This coding approach represents a form of content analysis. The relatively small number of codes helped us to reduce the recommendations to a manageable amount. For us, the coding also served as a process of indexing the text (Coffey and Atkins, 1996). In order to illustrate our analytical points on the non-COP15 recommendations we have illustrated our findings with fragments of text drawn from the recommendations.

Two distinct discussions: Lifestyle vs markets

Two ideological discussions were evident in the recommendations. The first discussion related to consumption and lifestyle, and the second discussion was concerned with markets and a critique of markets. Both themes are also reflected in contemporary academic literature on consumption (Campbell, 1998; Uusitalo, 2005; Jubas, 2007; Trentmann, 2009; Moisander et al, 2010), the first relating more to consumer policy and the second to international politics.

The discussion on consumption, lifestyle and education addresses a variety of issues relating to consumption in general, whereas the discussion on markets is more focused. The next sections show how these discussions emerged in the WWViews recommendations. Insights stemming from the recommendations will be further explored in the concluding section of this chapter.

Consumption, lifestyle and education

Consumption practices and lifestyles have been the subject of an enduring moral and practical discussion in Western countries after the rise of modern consumerism. In the Nordic countries there is a strong normative tradition questioning luxury and unnecessary consumption (Aléx, 2003). Similar moral suspicion of consumption has risen in the US and mainland Europe. Consumption based on needs (satisfaction) has often been considered to belong in the sphere of traditional society, whereas consumption based on wants (desire) is tied to modern society. This rhetoric relates to ideologies as well: the discussion on needs has its origins in puritan utilitarian philosophy and the discussion on wants within Romantic philosophy (Campbell, 1998).

In the WWViews recommendations, we reveal a demand for more sensible and rational consumption patterns which relate to normative and utilitarian discussion traditions. When these consumption practices are connected to environmental issues such as climate change, the moral and practical discussions tend to develop a new aspect where global responsibility and everyday practices are connected. Demands to 'consume less' are rather typical.

The participants emphasized sources and uses of knowledge. The conceptualization of knowledge suggests that local and other contexts shape consumption. Participants argued that in order to change consumption patterns, we need to draw on the knowledge that has been generated by consumers. The need for local knowledge and experience was articulated in the following forms: 'everyone should discuss what actions should be taken' (Brazil), 'to establish an educational program taking into account the characteristics of each region' (Bolivia) or 'invest in education related to climate change, taking into account the different realities' (Brazil) or to 'support all countries that are in the process of adaptation through local research activities, taking into consideration local, empirical knowledge' (Cameroon).

During the WWViews process, the distinction between needs and wants was not clearly drawn. Nevertheless, demands to consume less and educate people to use energy wisely can be seen as a subcategory of this discussion. Demands to limit consumption are connected to more traditional discussions about unselfish behaviour and consideration of the welfare of the nation, children and the planet. However, these recommendations are not limited by a national focus, as they clearly address a wider international audience.

Most of the recommendations suggest a transition towards sustainable consumption to replace current practices, which are viewed as wasteful and environmentally damaging. For instance, Belgium recommends low taxes (VAT) on sustainable products to encourage their production. There is little evidence in these demands that the participants would be advocating reduced consumption. Rather, they are advocating a change towards more environmentally efficient production that is sustainable. They are interpreting consumption as a quality question rather than a quantity question. A partial exception emerged from South Africa, where one recommendation advocated a slight reduction in consumption by calling for 'one day a month' to be declared a low carbon footprint day.

The more radical recommendations suggest either lower consumption or radical change in consumption practices, or both. Attention is given to manufacturing processes (e.g. in recommendations from Chile, Belgium and Finland) and the ways in which consumers themselves can affect the environment. Carbon footprint and environmental labelling are both common recommendations.

Vegetarianism was suggested as an effective strategy for emissions reduction in the recommendations from China-Taipei and US-Georgia, while Norway suggested strengthening family planning. China-Taipei also made a recommendation requiring global citizens to 'change lifestyle and reduce consumption'. While many of the recommendations are targeted at all countries or rich consumer societies, others articulate more local solutions. For example, participants in Indonesia-Jakarta recommended a 'local-based production and consumption model with environmental insights'. Most demands for sustainable consumption or even reduced consumption are targeted both at citizens and politicians. The recommendations tend to call for changes that ask for commitment from private citizens with the support of strong policy. For instance, a Finnish recommendation asks governments to 'guide consumption heavily in a more environmentally friendly direction'. The change required to deal with global warming is seen as so demanding and extensive that not only people's lifestyles, but also the entire economic and international system, are viewed critically.

A common consumer policy strategy to deliver product information is labelling (e.g. Thøgersen, 2005). This implies that the global community needs to find ways to gather and communicate information about carbon footprints in meaningful ways. Ideas for transparency included especially labels and standards that would make climate change information retrievable and accessible. A number of solutions were suggested as influential cues to reduce the cognitive requirements of buying (see Simon, 1990). These included 'permanent labels on products identifying and quantifying the levels of greenhouse gases emitted during manufacture and used to inform consumer choice' (Saint Lucia); indication of carbon footprints such that 'durability and reparability of products is taken into account' and guiding consumer's choices (Finland); and 'demand for more and uncensored information about the environmental impact of products, including labels on the quantity of energy used, carbon footprint and impact on ecosystem' (Chile). By increasing the transparency between processes and products, green labelling is seen to help everyday problem-solving such as choosing between products (Glaser, 1990). The transparency 'monitored and coordinated by international organizations' (Austria) would help everyone to make a contribution.

Demands for governments to inform public opinion were emphasized in the recommendations. Social transfers of knowledge were also proposed, such as 'aggressive awareness campaigning' (South Africa), 'promote awareness by incentive programs focused on reducing carbon emissions on a global scale' (US-Colorado), and 'tangible actions like advertising, media actions, posters, and gadgets are needed to make people aware' (Belgium-Flanders). The actions to manipulate public opinion were seen as desirable in terms of influencing citizens' climate change beliefs.

Local education and research activities can also be seen as new forms of participation and strategies used for political expression. Norris (1998) has argued that it is time to reinvent and reformulate participation and activism. Despite cause-oriented forms of action such as boycotts or demonstrations, there are now a growing number of organizations that serve as mediators of participation. Focusing on such organizations, the participants emphasized the links between consumer activism and the efforts made by the organizations.

Consumption is seen not as a matter of individuals but as a matter of community in all the recommendations categorized under consumption, lifestyle and education. Strong demands for more powerful policies were articulated, along with solidarity and a desire for sustainable consumerism. The WWViews participants suggested a strong model of global citizenship where consumer decisions are made by well-informed and solidarity-promoting individuals, where political leaders make decisions to support this desire, and where consumers have the tools to make sustainable decisions. While this ideal is strongly shared in developing countries, the rich industrialized countries were inclined to present more practical recommendations.

Markets and critique of markets

The level of trust in market mechanisms and international politics varies in the citizens' recommendations: many cases reflect a lack of trust in market mechanisms and, consequently, stronger policies are advocated. For instance, a recommendation from Austria suggests that new 'technologies should not be developed with a view to profits' and a recommendation from Bolivia points out a 'need to define national and international policies'.

However, there is no consensus on the role of markets. Some of the recommendations can be seen as strongly promoting market mechanisms, while others emphasize their weaknesses. In the pro-market recommendations, markets are seen as a way to create new technology and save the world, while sometimes even making a profit. For instance, a suggestion from US-Arizona depicts climate change as an opportunity for businesses, and recommends creating green private and public enterprises. However, pro-market recommendations are not suggesting that the market should be totally free; on the contrary, they see the state as a critical enabler of green markets. For example, a suggestion from Austria suggests that 'mandatory use and promotion of new technologies and economic and educational measures minimize global warming and create huge development opportunities for all countries'.

A demand for international solidarity, brotherhood and consciousness is a common request, raising the issue of international equality. Citizens in rich Western countries tend to make recommendations about their responsibility for helping poor countries. For instance, a recommendation from Norway suggests that the country must take a leading role by using its wealth to support new technology projects. Japan reminds us that we all share the same planet and we should 'gather our wisdom and hearts and implement possible commitments in each country'. Another Japanese recommendation asks that 'developed countries should take responsibilities as navigators, while developing countries with large emissions should build understanding and cooperation'.

Demands for corporations to reduce consumption are also common. While international models in politics and economics are questioned, some of the recommendations suggest a model based on local production and consumption. For example, Indonesia-Jakarta advocates both global responsibility as well as the local models mentioned before.

In the recommendations consumption decisions are not only seen as choices made by autonomous individuals, but also as results of collective processes that take into account a number of phases in collective decision-making, ranging from the identification of problems to deciding on solutions. Changes in consumption patterns were seen as a necessary precondition for stopping or slowing climate change. The focus on consumption processes can be traced back to Herbert Simon's idea of bounded rationality. Simon argues that when analysing a complex situation it is necessary to pay attention to the components that are likely to be taken into account by the actors (Simon, 1978). Components such as enhanced global awareness, public opinion or carbon footprint labels can be seen as a construction of the situation in which consumption takes place.

Although these processes are easily seen as patterns of individualistic consumers, they may also help consumers to restructure markets.

A critique of markets features more strongly in the recommendations than pro-market discussion. Citizens in industrialized and developing countries are equally critical of market mechanisms. In industrialized countries recommendations often focus on technical and financial solutions like '6 per cent VAT on sustainable products' (Belgium), 'Annex 1 countries are encouraged to show leadership through attainable, sustainable action designed to reduce emissions and invest in progressive market-driven technology and assist developing countries to do the same' (Canada) or 'open source development should be supported, which ensures free access by everyone to the developed technologies. Countries should invest in the fund a sum that corresponds to half of their military budget up until 2020' (Finland).

Discussion

Previously we have discussed policy recommendations contributed by citizens as part of the WWViews consultations. Our main focus has been on recommendations that address issues relating to both consumerism and citizenship. Based on our analysis, we argue that consumerism and citizenship are closely related in issues connected to climate policy. The consulted citizens formulated practical as well as political sugges-tions to change both the consumption and political culture.

Based on our analysis, we argue that citizens seek power from consumerism when lacking power in global political questions. This notion is also supported by research into the history and evolution of consumerism (Trentmann, 2006; Jubas, 2007). During recent decades there have been tensions between different rights connected to con-sumption, just as there have been in the fields of civic or human rights. Some arguments emphasize individual consumer rights, while others advocate cooperative models or public programmes to strengthen social networks or international trade (Jubas, 2007).

In the realm of global climate policy, the power of citizens is rather limited. The WWViews consultations demonstrated that participants responded to the focus on climate policy by also considering consumption and everyday matters. In this sense, the consulted citizens turned to consumer power. In doing so, the citizens saw that the realms of policy and private consumption are interconnected. The recommendations called for consumer policy to guide consumption through labelling and regulation, for instance. Moreover, education was called upon to inform people about consumer choices and support changes leading to more sustainable lifestyles. The role of educa-tion is therefore not only to teach consumers how to make wise choices or hold sellers accountable, but also to create shared cultural values where ideal lifestyles would be unselfish and respect the limits of natural resources. Recommendations concerning markets, on the other hand, were directed to political leaders rather than consumers. New organizations and a new kind of global citizenship were advocated.

The WWViews participants thus rejected the traditional dichotomy between selfish consumers and responsible citizens when afforded the opportunity to discuss climate matters that they had selected themselves (non-COP15 issues). These recommendations explore different kinds of consumerism and citizenship. Relationships between consumers and current policies are questioned, while demands for stronger policies are suggested. In these recommendations, the WWViews participants suggest that citizens and consumers share interests and goals. They also tend to think that all societies would benefit if people acted more responsibly and unselfishly, both as consumers and citizens. As argued before, this is not a new phenomenon but rather a strong tradition, which has historical precedents (Trentmann, 2006; Hilton, 2007). What is new here is that the global context is progressively replacing a narrowly defined national context.

Critiques of contemporary consumer lifestyle and consumerism contain practical ideas on how to develop new practices as well as new legislation to support sustainable consumption and a reduction in consumption. Tools for consumers to make better consumption choices are called for. Politics is here seen as an instrument that can build societies where good consumption is a possible choice for consumers. From a consumer policy perspective, it appears that this kind of consumerism could well be integrated into consumer policy (see Trumbull, 2006).

In a postmodern world, globalization and problems in international economy have increased demands to give more power to politics in general as well as strengthen citizens' opportunities to affect and participate in solving global issues, not only by electing national politicians, but also through global systems. Nevertheless, citizenship still retains its basis within the framework of the nation state. Many political scientists today emphasize the importance of global democracy as a key political issue of modern time (e.g. Paloheimo and Teivainen, 2004). The recommendations produced in the WWViews consultations promote a model of international democracy and strong politics. Such politics would support the everyday activities of responsible consumers and citizens alike.

Acknowledgements

Many thanks to translator Susan Heiskanen and editor Richard Worthington for revising the language. Further thanks go to anonymous reviewers.

References

Aléx, P. (2003) *Konsumera rätt – ett svenskt ideal: Behov, hushållning och konsumption*, [Consumer Rights – a Swedish Ideal: Need, Housekeeping and Consumption], Studentlitteratur, Lund, Sweden
Arnould, E. J. (2007) 'Should consumer citizens escape the market?', *The ANNALS of the American Academy of Political and Social Science*, vol 611, no 1, pp96–111

Barnett, C., Clarke, N., Cloke, P. and Malpass, A. (2005) 'The political ethics of consumerism', *Consumer Policy Review*, vol 15, no 2, pp45–51

Brewer, J. and Trentmann, F. (2006) 'Introduction: Space, time and value in consuming cultures', in J. Brewer and F. Trentmann (eds) *Consuming Cultures, Global Perspectives: Historical Trajectories, Transnational Exchanges*, Berg, Oxford

Campbell, C. (1998) 'Consumption and rhetorics of need and want', *Journal of Design History*, vol 11, no 3, pp235–246

Coffey, A. and Atkins, P. (1996) *Making Sense of Qualitative Data*, Sage, London

Glaser, R. (1990) 'Re-emergence of learning theory within instructional research', *American Psychologist*, vol 45, no 1, pp29–39

Helenius, R. (1974) *Konsumera allt och alla* [Consuming everything and everyone], Rabén & Sjögren, Stockholm

Hilton, M. (2007) 'Consumers and the state since the Second World War', *The ANNALS of the American Academy of Political and Social Science*, vol 611, no 1, pp66–81

Hirshman, A. O. (1970) *Exit, Voice, and Loyalty: Responses to Decline in Firms, Organizations, and States*, Harvard University Press, Cambridge, MA

Jubas, K. (2007) 'Conceptual confusion in democratic societies: Understandings and limitations of consumer-citizenship', *Journal of Consumer Culture*, vol 7, no 2, pp231–254

Keum, H., Devanathan, N., Deshpande, S., Nelson, M. R. and Shah, D. V. (2004) 'The citizen consumer: Media effects of the intersection of consumer and civic culture', *Political Communication*, vol 21, no 3, pp369–391

Lammi, M. (2006) 'Ett' varttuisi Suomenmaa. Suomalaisten kasvattaminen kulutusyhteiskuntaan kotimaisissa lyhytelokuvissa 1920–1969' [So Finland Would Grow. Raising Finns into Consumer Society in Domestic Short Films], PhD thesis, University of Helsinki, Helsinki

Livingstone, S., Lunt, P. and Miller, L. (2007) 'Citizens, consumers and the citizen-consumer: Articulating the interests at stake in media and communications regulation', *Discourse and Communication*, vol 1, no 1, pp85–111

Mayer, R. N. (1989) *The Consumer Movement: Guardians of the Marketplace*, Twayne, Boston

Micheletti, M. (2003) *Political Virtue and Shopping*, Palgrave Macmillan, New York

Moisander, J., Markkula, A. and Eräranta, K. (2010) 'Construction of consumer choice in the market: Challenges for environmental policy', *International Journal of Consumer Studies*, vol 34, no 1, pp73–79

Norris, P. (1998) *Democratic Phoenix*, Cambridge University Press, Cambridge

Paloheimo, H. and Teivainen, T. (2004) *A Possible World: Democratic Transformation of Global Institutions*, Zed Books, London

Pattie, C., Seyd, P. and Whiteley, P. (2003) 'Citizenship and civic engagement: Attitudes and behaviour in Britain', *Political Studies*, vol 51, no 3, pp443–468

Putnam, R. (1995) 'Bowling alone', *Journal of Democracy*, vol 6, no 1, pp65–78

Shah, D. V., McLeod, D. M., Friedland, L. and Nelson, M. R. (2007) 'The politics of consumption / the consumption of politics', *The ANNALS of the American Academy of Political and Social Science*, vol 611, no 1, pp6–15

Silverman, D. (2001) *Interpreting Qualitative Data: Methods For Analyzing Talk, Text and Interaction*, 2nd edn, Sage, London

Simon, H. A. (1978) 'Rationality as process and product of thought', *American Economic Review*, vol 68, no 2, pp1–16

Simon, H. A. (1990) 'Invariants of human behavior', *Annual Review of Psychology*, vol 41, no 1, pp1–19

Soper, K. (2007) 'Re-thinking the "good life": The citizenship dimension of consumer disaffection with consumerism', *Journal of Consumer Culture*, vol 7, no 2, pp205–228

Thøgersen, J. (2005) 'How may consumer policy empower consumers for sustainable lifestyles?' *Journal of Consumer Policy*, vol 28, no 2, pp143–177

Trentmann, F. (2006) 'The modern genealogy of the consumer: Meaning, identities and political synapses', in J. Brewer and F. Trentmann (eds) *Consuming Cultures, Global Perspectives: Historical Trajectories, Transnational Exchanges*, Berg, Oxford

Trentmann, F. (2008) *Free Trade Nation*, Oxford University Press, Oxford

Trentmann, F. (2009) 'Crossing divides: Consumption and globalization in history', *Journal of Consumer Culture*, vol 9, no 2, pp187–220

Trumbull, G. (2006) *Consumer Capitalism: Politics, Product Markets and Firm Strategy in France and Germany*, Cornell University Press, Ithaca, NY

Uusitalo, L. (2005) 'Consumers as citizens: Three approaches to collective problems', in K. Grünert and J. Thøgersen (eds) *Consumers, Policy and the Environment*, Springer, New York

van Deth, J. W. (2010) 'Is creative participation good for democracy?', in M. Micheletti and A. S. McFarland (eds) *Creative Participation: Responsibility Taking in the Political World*, Paradigm Publishers, Boulder, CO

Vidler, E. and Clarke, J. (2005) 'Creating citizen-consumers: New Labour and the remaking of public services', *Public Policy and Administration*, vol 20, no 2, pp19–37

Part IV
Discussing Cultural Variation and Local Engagement of Expertise

8

Representativeness and the Politics of Inclusion: Insights from WWViews Canada

Gwendolyn Blue, Jennifer Medlock
and Edna Einsiedel

All deliberative theorists specify that a process of political communication is not properly democratic unless it includes all persons affected by a decision; but I think that the meaning and conditions of this requirement are richer than most theorists acknowledge.

Iris Marion Young, 2006, pp49–50

In the past few decades, a range of initiatives has emerged that enables citizens to provide input into and potentially influence the policies that affect their lives. Referred to as deliberative and participatory processes, these initiatives redress the tendency to exclude citizens, their views and their knowledge from decision-making culture, particularly in relation to decisions that rely on scientific and technological expertise (Fischer, 2000; Bäckstrand, 2003; Jasanoff, 2003). More broadly, these initiatives are situated within the 'deliberative turn' in democratic theories and practices. Deliberative forms of democracy aim to renew public life by emphasizing the need for citizens to engage in public discussions in order to reach collective decisions about significant social issues (Dryzek, 1990, 2000; Barber, 1984; Young, 2000; Bohman, 2010).

While citizen participation in environment-related policy frequently takes place in particular locales and is often limited by the confines of the national state, the scope and complexity of environmental issues necessitates thinking in terms of global frameworks in addition to community-based and national ones. In this global context, attempts to orchestrate deliberative and participatory processes are challenging. Such initiatives call for approaches that simultaneously acknowledge and accommodate differences within and across regions (religion, culture, ethnicity and so forth), while at the same time they obliterate distance and difference in order to provide opportunities for dialogue, debate, and in certain instances, consensus. The ways in which these tensions are negotiated in practical settings can have significant bearing on the wielding of voice in public debates as well as the kinds of perspectives about the environment that will inform emerging global policies.

WWViews is the first instance of a global public formed with the specific remit to deliver citizen perspectives to United Nations climate policy negotiations. Combining face-to-face citizen meetings in each participating nation with a web interface to compare results across nations, WWViews provides a unique and timely forum for public debate on a global scale. It also affords the opportunity to reflect on the practical and theoretical implications of convening global publics to discuss global environmental policy.

The intent of this chapter is to examine the ways in which the WWViews process accommodated differences across social groups while it also created the conditions for a shared public conversation at a global level. As organizers and researchers situated within a university context, we bring critical theoretical debates to bear on the ways in which we selected our participants and organized our event. By 'critical theory', we mean scholarship that is concerned with charting, understanding and transforming oppressive social forces that serve to privilege certain social groups at the expense of others (Dryzek, 2000, p20). While in general the objective of most deliberative processes is to address conditions of inequality by providing opportunities for a range of people to participate, all too often these processes are grounded in the tacit assumption that, once at the table, citizens are similar to one another rather than characterized by profound social and cultural differences, some of which are marked by deeply rooted inequalities. If conditions of inequality among social groups are not taken into account, deliberative initiatives risk reinforcing and perpetuating social hierarchies, a point long argued by feminist critics of deliberative democratic initiatives. A central task of a theoretically informed practice is to render visible the ways in which social inequalities pervade social interaction, and to envision ways to address these inequalities.

In what follows, we compare the selection process promoted by the Danish Board of Technology (DBT) with the practices deployed by WWViews Canada. First, we discuss a fundamental tension at play in WWViews between the need to establish a legitimate cross-section of the population of participating nations, on one hand, and the need to acknowledge and account for salient differences between groups, on the other. Although the DBT emphasized the criterion of representativeness in its methods for participant selection, the Canadian organizing team attended more centrally to what political theorist Iris Young (2000) refers to as the politics of inclusion. Next, we outline the procedures we put in place to deepen and extend our inclusive practices, with specific attention to Canada's political, cultural and social context. In an effort to bridge our theoretical insights with our organizing strategies, we implemented the following measures: expanding the random selection process to target marginalized groups with a particular emphasis on northern and Aboriginal populations, incorporating public rituals of acknowledgement, and creating separate spaces for deliberation. To conclude, we discuss the implications and challenges of incorporating inclusive procedures at national and international levels. In planning future events, we argue that it is important to make assumptions about inclusion and representation explicit as these can enable and constrain the transformative potential of attempts to democratize environmental policy-making cultures.

From the criterion of representativeness to the politics of inclusion

The overarching normative goal of WWViews is one shared by deliberative democracy initiatives in general: collective decisions should be subject to public debate where a range of people and perspectives, rather than a small elite group with vested interests, determine the course. A fundamental guiding assumption is that ordinary citizens (or 'non-experts') have the capacity to engage seriously and intelligently with complex issues and decisions that will affect their lives.

One of the central challenges in organizing public deliberation events lies with how to ensure the representation of a range of people while also addressing complex issues of inclusion. These challenges are amplified in large-scale events. For example, many theorists and practitioners assume that deliberation should occur in a single forum where deliberators speak directly to one another and where face-to-face interaction is a prerequisite for deliberative democracy (Mansbridge, 1980; Barber, 1984; Dryzek, 1990). Beyond local contexts, however, geographical distance and the numbers of participants limit the practicality of face-to-face communication. Others adapt deliberative processes by combining face-to-face communication with survey questions and polling techniques to solicit responses from a number of people located across large geographic distances (Fishkin, 1999). In a similar fashion, WWViews combined DBT's expertise in small-scale consensus conferences rooted in deliberation and consensus-building with a series of voting questions that served as a form of global public opinion poll. The project design, according to DBT, 'makes it potentially possible for all nations on Earth to take part and to produce comparable results that can be clearly communicated to policy-makers' (Bedsted and Klüver, 2009, p3).

In organizing this global event, a series of trade-offs had to be negotiated between ensuring a politically legitimate sample of the population, on one hand, and enabling an affordable process that recruits manageable numbers to engage in discussion and deliberation, on the other. The guidelines stated that each participating nation recruit approximately 100 citizens. DBT was careful to note that at a national level, these numbers are not large enough to ensure statistical representation because of the limited number of citizens per event; however, on a global scale, the process provided significant numbers to give an indication of global attitudes and opinions (Bedsted and Klüver, 2009, p6). To achieve a politically legitimate cross-section of the global population, one of the methodological aims of DBT was to ensure that the selection process used by each participating nation reflected its demographic distribution. Guidelines for the selection of participants were put in place to ensure the reliability of the final results. To the degree possible, the selection of participants reflected the demographic distribution of each nation or region based on age, gender, occupation, education and geographical residence. Where appropriate, additional demographic criteria such as race and ethnicity could be incorporated. One limiting criterion was that participants could not hold positions of decision-making or expertise on climate change nor were they to be stakeholders representing particular interest groups. These guidelines were communicated to organizers in each nation in a preliminary training workshop as well as through instructional manuals.

In setting these criteria, DBT aimed to meet the criterion of representativeness shared by most public participation initiatives. This refers to the condition that participants should comprise a representative sample of the broader public, rather than a self-selected subset (Rowe and Frewer, 2004, pp12–13). Theoretically, democratic processes demand that all members of affected communities, or at least their representatives, be consulted and involved in decision-making processes. Methodologically, representativeness is important if the aim is to gauge the opinions of the general public. Politically, the appearance of a sampling bias can undermine the credibility of the initiative. Although representativeness is an important criterion for approaching a legitimate sample of any population, it is a complicated ideal that exists in tension with other democratic criteria, particularly those rooted in deliberative understandings of democracy. Practical constraints, for example, must be taken into consideration. In order to meet the criterion of representativeness for all affected populations, large samples of people are required. This condition, however, curtails the potential for sustained and substantial discussion, deliberation and debate.

The criterion of representativeness can also lead to the assumption that social differences have been accounted for in deliberative processes. As discussed, the normative ideal underlying events such as WWViews is that those affected by a decision should be included in the decision-making processes and should have had the opportunity to influence the outcome. Yet, as the epigraph to this chapter suggests, including people in decision-making processes is nuanced, complicated and warrants sustained critical reflection before deliberative processes commence, as well as after they have concluded. The criterion of representativeness should be considered a necessary starting point for participant selection, but not an end in itself. It is also important to address differences among social groups, particularly differences that may play a role in limiting the participation of some groups during deliberative events.

Feminist critiques of deliberative processes have pointed to the underlying assumption that it is possible to deliberate as equals in the face of underlying structural inequalities (Young, 1989, 2000; Fraser, 1990; Benhabib, 2002). Social inequalities are never fully removed; they are simply bracketed. Practically, status inequalities remain, embedded in protocols of communication, style and decorum, preventing participants from engaging as equal peers. Just because deliberative arenas are deemed to be spaces where social distinctions are neutralized does not make them, in practice, innocuous sites. Iris Young argues that deliberative processes are inclusive to the extent that they attend to the social relations that position people and condition their experiences, opportunities and knowledge (2000, p83). According to Young, participatory democratic practices must be structured starting from the assumption that some social groups are oppressed or disadvantaged in relation to others. This means taking into account the general historical and social conditions in which we are situated. A democratic public, in other words, arrives at efficacious political judgements not by ignoring or bracketing social differences but by accounting for and communicating the perspectives that have been conditioned by them.

Here, it is important to distinguish a 'social group' from a demographic category. The criterion of representativeness is based on aggregate notions of populations in which people are categorized by certain attributes held in common such as age, income, geographical location and so forth. These categories do not account for more intractable, but no less significant, issues of identification, difference and inequality.[1] Unlike demographic categories that are often arbitrarily chosen for statistical purposes, social groups are united by a sense of common identity, values, language, history and so on. Social groups, in other words, are integrated in particular cultural and social contexts. In certain instances, social groups are differentiated from others by virtue of the ways in which they have been oppressed by structural conditions of exploitation, marginalization or powerlessness in the face of violence and harassment (Young, 1989, p261). One reason to call these processes structural is that they are relatively permanent. Although the specific content and detail of these relations are dynamic and evolving, social group differences are also reproduced, passed on through generations via assumptions, beliefs and practices. It is important to note that most people identify with multiple social groups, an identification that changes over time and in relation to particular contexts. The degree of oppression or marginalization of a group also changes depending on its context. As such, one challenge in accounting for social group differentiation lies with deciding which groups deserve special recognition in deliberative processes.

The criterion of inclusion encourages attention to both the explicit and tacit ways in which certain social groups can be excluded from deliberations. As Young elaborates (2000, pp53–54), exclusions from political decision-making can be external or internal. External exclusions concern how certain individuals and groups are purposely or inadvertently left out of public discussions and decision-making. By making issues of transparency, accountability and access to deliberative publics central to their concerns, most theorists and practitioners are attentive and responsive to mechanisms of external exclusion.

Less acknowledged, however, are situations concerning internal exclusionary mechanisms. These refer to the myriad ways in which people lack effective opportunities to participate in meaningful dialogue and influence the thinking of others, even though they may have access to decision-making forums and procedures. As internal exclusionary mechanisms comprise the cultural fabric that engulfs daily life, they often elude theorists and practitioners. We carry with us particular traditions and assumptions in the ways we eat, dress and communicate: these practices are culturally particular and might unintentionally exclude or alienate people who do not share similar cultural backgrounds. For example, an over-reliance on certain styles of expression as rational, dispassionate and orderly can have exclusionary implications, particularly for those who belong to groups in which emotional and tangential modes of expression are central modes of communication. Events as seemingly innocuous or supplementary such as rest breaks or meal options can have implications for the ways in which participants feel comfortable with deliberative processes. Moreover, certain exclusions have become widely accepted as normal and legitimate at the level of the nation state even though they can have dire exclusionary consequences. Many people have been deprived of lands, livelihood, as well as cultural and religious practices by colonial and post-colonial systems of appropriation and exploitation, placing them in precarious and vulnerable positions.

To summarize, in addition to accounting for the criterion of representativeness to ensure a legitimate deliberative process, it is also important to attend to the multifaceted differences among social groups, particularly in instances where certain groups are at a disadvantage due to broader structures of social inequality. By attending to both external and internal mechanisms, organizers can work towards ensuring that participatory events are more inclusive of a range of social groups. If we rely solely on demographic categories and external mechanisms for inclusion, we risk ignoring, and at worst perpetuating, hierarchies of social power.

WWViews Canada: Accounting for the norm of inclusion

One of our primary concerns in organizing WWViews Canada involved implementing inclusive mechanisms for northern and Aboriginal participants.[2] As many scientists have illustrated, climate change is experienced earlier and more acutely in polar regions than in other parts of Canada and disproportionately affects Aboriginal populations (Kattsov and Kallen, 2005; Ford et al, 2006). Complex legacies of colonization have structurally disadvantaged Aboriginal peoples across Canada, contributing to endemic social and environmental problems in these communities. Given the greater risks and vulnerabilities imposed by climate change on northern and Aboriginal communities, we felt it was important to emphasize the perspectives of these populations and to ensure that we were not tacitly excluding them from the deliberative process.

Previous large-scale deliberative initiatives in Canada have failed to acknowledge the significant power relations that marginalize Aboriginal communities, preventing them from expressing valuable perspectives in deliberations. Von Lieres and Kahane's (2007) analysis of the Canadian Romanow Commission provides insights into the often well-meaning but nevertheless exclusionary practices that can unfold in public deliberations. Established in 2001 by the Canadian government to deliberate with citizens on the future of healthcare, the mandate of the Romanow Commission was to engage Canadians in a discussion about the future of Canada's healthcare system.

Consultations took many forms: televised forums, open public hearings, consultation workbooks as well as a National Citizens' Dialogue. The intent of the National Dialogue was to take citizen involvement beyond consultation to deliberation: that is, to engage citizens in dialogues that challenge their existing understandings, confront them with trade-offs and encourage them to define paths to reform that fit with their deepest values. The Commission framed the participants in terms of statistically representative groups rather than recognizing the structural inequalities among groups. Although Aboriginal peoples were included in deliberations, the discussion questions did not focus on Aboriginal health per se, nor were mechanisms built in to deal with issues of marginalization and disenfranchisement that might arise in the process of the deliberation. Inclusion of Aboriginal peoples, in other words, was more of a token gesture than an attempt to radically transform health policy to suit the needs of Aboriginal populations. In recognizing the limitations of deliberative processes such as the Romanow Commission with respect to Aboriginal populations, we sought mechanisms to avoid reproducing these in our own deliberations. Although we were mindful of the structure set forth by DBT, we had a degree of latitude to customize the process to account for the structural inequalities inherent in our national context. We addressed the norm of inclusion in the following ways: by expanding the random selection process to target marginalized groups with a particular emphasis on northern and Aboriginal populations; by incorporating public rituals of acknowledgement; and by creating separate spaces for deliberation. We discuss each of these in turn, providing details regarding our specific methods for participant selection.

Random and targeted selection processes

Our selection of participants used a combination of random selection and targeted recruiting. Initially, we hired a national research firm to provide a list of 5000 names based on a selection criteria of adult Canadians (18 and older) that accounted for provincial population figures, as well as age, ethnicity, gender and income distribution. From this list, letters of invitation were sent to 3000 Canadians. The letters of invitation introduced the WWViews project and described the role of the citizen panellists. We emphasized that citizens did not have to have previous knowledge of climate change or climate-related policy issues. Interested participants were asked to complete an online application with a series of questions regarding age, gender, household income, occupation, province of residence, citizenship status[3] as well as an optional question on whether the applicant was a member of a visible minority or an Aboriginal population. Participants were also asked to provide a short statement on why they were interested in being part of the panel. Options were available to submit applications via fax or post for those who lacked access to computer technology.

Due to particular geographical constraints, the Canadian population is not distributed uniformly.[4] Most people inhabit the warmer southern areas, leaving northern areas largely uninhabited. The largest concentrations are located in the eastern provinces of

Quebec and Ontario and in the large metropolitan centres of Toronto, Montreal and Vancouver. With demographic representation as our sole guiding framework, we would have had significantly more representation from urban areas than rural areas, and from southern Canada than from northern communities. This would prevent us from providing adequate places for those who suffer the immediate consequences of severe climatic changes, as discussed previously.

Although only 3.8 per cent of the Canadian population identifies as Aboriginal, we allocated ten spots (or 10 per cent representation) to Aboriginal participants with an emphasis on persons from northern territories. To compensate for demographic distribution by population, spots were taken from the participant allocation of the two largest Canadian provinces by population, Ontario and Quebec. We targeted Aboriginal populations by sending 'calls for participants' to a range of organizations such as the Inuit Circumpolar Institute as well as various community and friendship centres. We also made use of public service announcements in local radio stations and newspapers. In addition to targeted recruitments for Aboriginal participants, we also contacted a wide variety of community groups that work with immigrant populations (e.g. Alberta Network of Immigrant Women, Calgary Catholic Immigration Society). To increase rural representation, we contacted farmer organizations and other groups affiliated with rural communities (e.g. 4-H groups, Reduced Tillage Linkages, Farmers' Market Association, etc.).[5]

Public rituals of acknowledgement

While explicit targeting of marginalized populations addresses external modes of exclusion, it does not account for more intangible internal exclusionary mechanisms. As discussed, these mechanisms are more cultural than procedural in form. In attending to these internal mechanisms, we accounted for what Young refers to as 'public greetings' or 'rituals of acknowledgement'. These refer to gestures of respect and politeness that announce the willingness to listen and take responsibility for the group as a whole, while at the same time acknowledging each person and group in their particularity. As Young describes:

> *Actual political discussion should be examined not only for what it says, whether the issues are well formulated, the arguments coherent and so on. We should also ask whether major contributions to a political debate show discursive signs that they are addressing all those who should be included in the debate.* (2000, p62)

Although they might appear to be frills or extraneous decorations to the deliberative business at hand, rituals of acknowledgement are extremely important for setting a context of respect and recognition, particularly for marginalized groups. In some cases, they can be instrumental in keeping difficult discussions on track. Political meetings, for example, often begin with receptions in which participants greet one another personally. These are attempts to symbolically recognize others as included in the discussion – especially those who differ in social location, opinion and interest.

Our organizing team spent a great deal of time discussing and debating how to foster a more inclusive climate. In recognition of our Aboriginal participants as well as the fact that our event was held in an area with competing jurisdictional claims, we included a greeting from an official government representative from the City of Calgary[6] as well as a prayer and blessing from a First Nations Elder. At our Saturday evening dinner, three young female Inuit participants demonstrated traditional Inuit throat singing for the entire group. We also had an Aboriginal volunteer who met our Aboriginal participants at the airport and was available throughout the event to answer questions and provide support where necessary. To 'bring' the Canadian Arctic into our discussions, our keynote speaker journalist Ed Struzik focused his presentation on his adventures in the Arctic, drawing attention to the ways in which climate change was transforming landscapes as well as human and non-human lives.

Language disparities comprised another challenge we faced in facilitating a climate of inclusion. Although the majority of the Canadian population is anglophone (58 per cent), a significant portion (22 per cent) of the population is francophone. Canada is also officially a bilingual nation. As such, our deliberations were held in both English and French, although in order to address logistical difficulties as well as reduce the significant costs of holding a bilingual deliberation, we needed to make concessions where possible. All our publications and correspondence, for example, appeared in both official languages. Rather than conducting the entire deliberation in both languages, we allocated two 'francophone' tables, with French-speaking facilitators and note-takers. The facilitators were responsible for translating the moderator's comments into French. The deliberations, voting and citizen recommendations were conducted in French and later translated back into English. Although the moderators' abilities in French were limited, the francophone participants greatly appreciated when attempts were made, however clumsily, to address the entire group in French. With hindsight we would have made more effort to deliver the moderators' address in both languages.

Separate spaces for deliberation

Perhaps one of our most difficult challenges revolved around whether to create separate spaces for deliberation or to integrate the participants with one another. On the whole, we decided that participants would be assigned to different tables for the Friday afternoon, Saturday morning and Saturday afternoon sessions. In doing so, we wanted to ensure that participants had a chance to meet and dialogue with a range of different people rather than remaining with the same group for the duration of the event.

Given our emphasis on creating inclusive spaces for Aboriginal and northern participants, we proposed allocating a separate table for these participants to create a more conducive environment for deliberation. Our objective was to recognize distinct cultural and structural group differences, as we had done for our francophone participants.

This would mean that, unlike the other participants, these groups would remain together for the duration of the deliberation. We were concerned, however, that such a move might further marginalize our Aboriginal participants, preventing them from integrating and sharing their insights with the broader group. In making our final decision, we consulted with our Advisory Committee. We also consulted with our Aboriginal participants about their preferences. Seven participants requested to remain at the same table for the duration, whereas three preferred to circulate with the broader group. We also hired a facilitator from a local First Nations community to guide the discussions at the Aboriginal table.

As an organizing team, we negotiated tensions between the criterion of representativeness promoted by DBT and attention to the politics of inclusion by addressing the differences and inequalities among social groups. Our aim is to draw attention to the theoretical arguments for taking into account social differences as well as to illustrate strategies that can be incorporated to bring these insights into practical contexts. While we were attentive to the structural inequalities with respect to our Aboriginal and northern participants, as well as the language differences among anglophones and francophones, outside of targeted recruiting techniques we were not able to compensate for other social differences based on ethnicity, class or rural geographical location, nor did we make concessions for the participation of non-citizens.

Concluding remarks

As a response to as well as a reflection of the ways in which scientific and environment-related policies are increasingly pursued on a global scale, WWViews provides a timely experiment in forging global spaces for public dialogue that simultaneously work within as well as transcend national borders. Given that environmental policy-making reveals emergent aspects of transnational politics that will grow in significance this century, global deliberative initiatives are significant sites for theorizing and implementing visions of transnational democratic practices, particularly as new political forums emerge in which lay publics play a more significant role in decision-making cultures.

A core tension permeating these initiatives lies between the universalizing and unifying forces of globalization, on one hand, and the nuanced, particular and multiple forms of local knowledge claims and identities, on the other. As political theorist James Bohman (2010) argues, democracy needs to be rethought according to a new transnational ideal in which we move from a notion of democracy as *demos*, rule by the people as dictated by territorial identification, to one in which we foster a pluralistic understanding of *demoi*, rule by peoples across national borders. In doing so, however, it is important to be mindful of the deeply rooted structural inequalities that persist within and between nations, and that are unlikely to disappear within emergent global frameworks without significant attention, consideration and intervention.

Our intent in this chapter has been to elaborate on the ways in which we negotiated such tensions in the practical context of organizing the Canadian section of WWViews. In doing so, we illustrate some strategies by which the local dynamics of exclusion and marginalization can be managed within a global deliberation process. While aware of the need to ensure consistency across nations in the selection process, we are also attentive to the importance of accounting for salient differences among social groups in Canada. In order to address the politics of inclusion that are often eclipsed by the criterion of representativeness, we incorporated the following mechanisms: targeted sampling, rituals of acknowledgement, and separate spaces for deliberation. By enabling different opportunities for Aboriginal participants, we aimed to address, albeit insufficiently and inadequately, the histories of colonization and exploitation that have resulted in the decline of Aboriginal populations due to targeted genocide campaigns as well as forced relocations and appropriation of lands. Moreover, we acknowledge that these communities are more vulnerable to the consequences of climate change – a condition that urgently warrants the inclusion of their voices in climate policy decision-making.

Our attempts to create a more inclusive process were limited in two significant ways. First, it is important to recognize that our strategies for attending to the politics of inclusion come with their own unavoidable trade-offs and tensions. In creating separate spaces for some Aboriginal participants, for example, we separated their perspectives from the larger group, since time constraints prevented us from creating opportunities to share information among tables. In other words, creating separate spaces for deliberation (separation) came at the cost of sharing views among participants (integration). Second, our efforts were limited in large part by the predetermined structure of WWViews. As the questions for discussion were set in advance, we did not have latitude to incorporate insights from our participants into the topics for discussion during the deliberation. Our participants did not have the opportunity to question the agenda set out for them, as well as the implicit values and assumptions that structured the deliberation. Since different social locations can provide distinct understandings and unique perspectives on social issues, it is important to incorporate these into the broader debates. Unfortunately, we did not have mechanisms to share these with the entire group (nationally and internationally), thereby curtailing the transformative potential of the deliberative process. This acknowledged problem at the level of the nation state is exacerbated at international level. Beyond a very superficial level of engagement, participants were not given insights into the profound structural differences among nations, particularly the ways in which these play out in everyday settings. Although comparisons between nations were possible via the web interface, these comparisons did not indicate the differences within and between nations that might have enriched the discussions at hand.

In drawing attention to differences among social groups, and in reflecting on how these differences might be incorporated into public deliberation initiatives such as WWViews, we do not offer definitive solutions. Rather, we hope to encourage further discussion, debate and research. Attending to how to incorporate social differences into global dialogue events, while at the same time ensuring that the process is easy and affordable to implement, will be a challenge for DBT in organizing future events. The focus on 'comparable results' that can be 'clearly communicated to policy-makers' might serve to undermine efforts to ensure that differences within and among nations are taken into account (Bedsted and Klüver, 2009, p3). Values such as clarity, commensurability and consensus exist in tension with deliberative ideals that aim to incorporate a range of perspectives, beliefs and attitudes, as these ideals can often result in complexity, heterogeneity and conflict. While the latter qualities do not lend themselves to sending clear and concise messages to policy-makers, they do enable a forum in which differences among groups can be articulated, shared and potentially incorporated into decision-making processes. In accounting for differences among social groups, an evaluative challenge is also at hand: how can we determine whether the inclusive mechanisms incorporated are successful? What criteria would define 'success' in these instances?

Theoretically, public participation events have potential to reshape existing power relations as they provide mechanisms to include a range of people and perspectives in decision-making processes, particularly those who have been previously marginalized. Yet, as feminist critiques have long emphasized, the normative aims of public participation are difficult to implement in a world with structural inequalities. Successful strategies must confront not only the external manifestations of democratic practices, but also the internal cultural logics, the quotidian practices and social relations that constitute, reproduce and make possible the dominance of exclusionary practices. The effects of neglecting the cultural and social politics of deliberative initiatives are far from benign. Structural inequalities between groups cannot simply be bracketed in the attempt to create the appearance of a politically legitimate sample of a population. Attempts to foster 'neutrality' in the selection process are themselves political as they can mask the structural inequalities that inform and guide contemporary practices. As Iris Young states, attempts to repoliticize public life should not be focused on creating a unified public realm 'in which citizens leave behind their particular group affiliations, histories and needs to discuss a general interest or common good' (2000, p258; see also Hayward 2008). Rather, practitioners should aim to create heterogeneous publics in which social differences are recognized, acknowledged and addressed. The fundamental challenge lies with how to bridge these theoretical insights with practical, affordable and manageable strategies for implementation.

Acknowledgements

This research was supported by a research grant to the first author and a public outreach grant to the last author from the Social Sciences and Humanities Research Council's Canadian Environmental Issues Program. The WWViews-Canada public consultation was funded by the latter grant.

Notes

1   In acknowledging the limitations of demographic categories, we do not intend to dismiss their relevance. Most forms of social science, after all, are dependent on these categories; it is difficult, if not impossible, to talk about society or politics without them (Christians and Carey, 1989). Our intent is to take responsibility for how these categories are used, and how they relate to the fostering or inhibiting of democratic practices.

2   Aboriginal is an umbrella term for a culturally, linguistically, economically, socially and politically diverse set of communities and nations including First Nations, Inuit and Metis.

3   The project team made the decision to include only Canadian citizens in the consultation. While we were attentive to the exclusionary implications of this choice, our rationale was based on the realization that the aim of the consultation was to provide advice to elected policy-makers who are ultimately responsible to their constituents.

4   Demographic composition, based on 2006 census data (Statistics Canada, 2008): 81 per cent of Canadians live in a metropolitan area, with the majority living in Montreal, Toronto or Vancouver; largest visible minority groups are Chinese and South Asian; persons reporting Aboriginal identity represented 3.8 per cent of Canadian population. Of this, the majority (60 per cent) are North American Indians, 34 per cent are Metis and 4 per cent are Inuit.

5   In total 98 applications were received from the random selection recruitment process and 8 were received through targeted methods of recruitment; 3 dropped out in advance of the event, meaning that 103 participants attended the final event. In total, we recruited 10 Aboriginal participants (2 through random selection, 8 through targeted selection).

6   It is not insignificant that Federal Environment Minister Jim Prentice was absent from delivering these greetings, even though his political constituency is in Calgary. Although we made numerous attempts to contact his office and to provide avenues for addressing our participants, our requests were repeatedly ignored or denied until well after the consultation had taken place. This signals a lack of timely acknowledgement by our national government of the importance of citizen input and by extension of the democratic process.

References

Bäckstrand, K. (2003) 'Civic science for sustainability: Reframing the role of experts, policy-makers and citizens in environmental governance', *Global Environmental Politics*, vol 3, no 4, pp24–41

Barber, B. R. (1984) *Strong Democracy*, University of California Press, Berkeley, CA

Bedsted, B. and Klüver, L. (2009) *World Wide Views on Global Warming: From the World's Citizens to the Climate Policy-makers: Policy Report*, Danish Board of Technology, Copenhagen

Benhabib, S. (2002) *The Claims of Culture: Equality and Diversity in the Global Era*, Princeton University Press, Princeton, NJ

Bohman, J. (2010) *Democracy Across Borders: From* Demos *to* Demoi, MIT Press, Cambridge, MA

Christians, C. and Carey, J. (1989) 'The logic and aims of qualitative research', in G. Stempel and B. Westley (eds) *Research Methods in Mass Communication*, Prentice Hall, Englewood Cliffs, NJ

Dryzek, J. (1990) *Discursive Democracy: Politics, Policy and Political Science*, Cambridge University Press, Cambridge

Dryzek, J. (2000) 'Liberal democracy and the critical alternative', in J. S. Dryzek (ed) *Deliberative Democracy and Beyond: Liberals, Critics, Contestations*, Oxford University Press, Oxford

Fischer, F. (2000) *Citizens, Experts and the Environment: The Politics of Local Knowledge*, Duke University Press, Durham

Fishkin, J. (1999) 'Toward deliberative democracy: Experimenting with an ideal', in S. L. Elkin and K. E. Soltan (eds) *Citizen Competence and Democratic Institutions*, Pennsylvania State University Press, University Park, PA

Ford, J., Smit, B. and Wandel, J. (2006) 'Vulnerability to climate change in the Arctic: A case study from Arctic Bay, Canada', *Global Environmental Change*, vol 16, no 2, pp145–160

Fraser, N. (1990) 'Rethinking the public sphere: A contribution to the critique of actually existing democracy', *Social Text*, vol 25/26, pp56–80

Hayward, B. (2008) 'Let's talk about the weather: Decentering democratic debate about climate change', *Hypatia*, vol 23, no 3, pp79–98

Jasanoff, S. (2003) 'Technologies of humility: Citizen participation in governing science', *Minerva*, vol 41, no 3, pp223–244

Kattsov, V. and Kallen, E. (2005) 'Future climate change: Modeling and scenarios for the Arctic', in *Arctic Climate Impact Assessment: Scientific Report*, Cambridge University Press, Cambridge, UK

Mansbridge, J. (1980) *Beyond Adversary Democracy*, Basic Books, New York

Rowe, G. and Frewer, L. (2004) 'Evaluating public participation exercises: A research agenda', *Science, Technology & Human Values*, vol 29, no 4, pp512–556

von Lieres, B. and Kahane, D. (2007) 'Inclusion and representation in democratic deliberations: Lessons from Canada's Romanow Commission', in A. Cornwall and V. Coelho (eds) *Spaces for Change: The Politics of Citizen Participation in New Democratic Arenas*, Zed Books, London

Statistics Canada (2008) *Profile of Ethnic Origin and Visible Minorities for Urban Areas, 2006 Census*, www.statcan.gc.ca/bsolc/olc-cel/olc-cel?catno=94-580-X2006009&lang=eng, accessed 20 April 2011

Young, I. M. (1989) 'Polity and group difference: A critique of the ideal of universal citizenship', *Ethics*, vol 99, no 2, pp250–274

Young, I. M. (2000) *Inclusion and Democracy*, Oxford University Press, Oxford

Young, I. M. (2006) 'De-centering deliberative democracy', *Kettering Review*, vol 24, no 3, pp43–60

9

WWViews in the Global South: Evidence from India

Ravtosh Bal

The idea of global deliberation raises important questions about how cultural factors impact the travel of standardized methods. Can a uniform deliberation model based on the Habermasian public sphere be adopted in all contexts, irrespective of the local culture, systems of governance and styles of decision-making? Some previous studies on the consensus conference model suggest that the participatory deliberation model 'travels well' (Einsiedel et al, 2001) and can be applied in different national contexts.

Yet other studies claim that the national context determines the effectiveness of the model. Nielsen et al (2007), for example, look at three consensus conferences on genetically modified organisms (GMOs), which took place in France, Norway and Denmark. The authors argue that the concept of participation embodies values that vary from country to country and is subject to differing interpretations leading to distinct ideas about what constitute legitimate goals for participatory processes. Similarly, Dryzek and Tucker (2008), in a comparative study of consensus conferences on the issue of genetically modified food in Denmark, France and the US, claim that the type of political system has an important influence on the potential of deliberations and their policy impact. Others (see Bhargava and Reifeld, 2005) have argued that the concept of the Habermasian public sphere is one that has its genesis in a particular period and culture and may exist, if at all, in a very different form in other cultures.

In this chapter, I focus on the manner in which the WWViews model was applied in India, which is a large and diverse democracy that is also characterized by the existence of deep vertical and horizontal inequalities. India, along with many other non-Western countries, has a long tradition of democracy and deliberation. Amartya Sen (2005), for example, has detailed the long history of what he terms the argumentative tradition in India, which has allowed an acceptance of diverse viewpoints and dialogues. The role of reasoning is emphasized in Sen's (2009) idea of social justice, referring to 'government by discussion' in the context of democracy. According to Sen, democracy as public reasoning is dependent on dialogue and argumentation and is a critical instrument to deal with social inequalities:

> *the critical voice is the traditional ally of the aggrieved, and participation in argu-*
> *ments is a general opportunity, not a particularly specialized skill (like composing*
> *sonnets or performing trapeze acts).* (Sen, 2005, pxiii)

Social justice, in this view, requires more democratic participation, and public reasoning combined with democratic politics is the way to go. The criterion of inclusion, in this way of thinking, becomes a central concept when reflecting and evaluating the role of public consultations and deliberation exercises, especially in culturally diverse, multilingual and stratified societies such as India.

I will study how the WWViews process was organized in India, evaluate its levels of inclusiveness and discuss the challenges related to the adaptation of the deliberation model in the hierarchic context of Indian science and technology (S&T) policy. The WWViews event in India was organized at two locations, New Delhi and Bangalore. Both are large urban conglomerates: the former is the political capital of the country, while the latter is its IT capital. The purpose of the analysis is to understand where deliberation fits in the current science policy scenario and how to make such events more inclusive in the future. Bhargava (2005, p46) makes this point in the context of outlining a task for an ethical political theory but it is equally a goal for practitioners:

> *Public deliberation is better than violent-prone demonstrations, if and only if agents*
> *have a real choice between these two options. It is futile, irresponsible and even un-*
> *ethical to ask people to deliberate when neither the opportunity nor the capacity for*
> *deliberation exists. However, it is equally irresponsible not to struggle to transcend*
> *this situation and build conditions for a more inclusive public deliberation.*

As we shall see, the experience of WWViews in India is particularly instructive in identifying some of the procedural and contextual factors that can inhibit inclusive deliberation.

Theory and methods

Theoretically I draw upon the traditions in policy sciences and in science, technology and society (STS) studies that deal with participatory governance and technology assessment. Both traditions focus on public participation through the concept of 'deliberation'. According to Burkhalter et al (2002, p418), face-to-face deliberation refers to:

> *(a) a process that involves the careful weighing of information and views, (b) an*
> *egalitarian process with adequate speaking opportunities and attentive listening by*
> *participants, and (c) dialogue that bridges differences among participants' diverse*
> *ways of speaking and knowing.*

According to this definition, people deliberate when they jointly examine a problem by respectfully considering the viewpoints of others in an inclusive environment and arrive at well-reasoned recommendations.

The benefits of participative, deliberative exercises have been highlighted by practitioners as well as theorists. On the one hand, many scholars argue for participation as an end in itself that gives meaning to democracy (Fischer, 2000). Others (e.g. DeLeon, 1997) emphasize the effectiveness of policies that result from a democratic discourse reflecting the values and preferences among various stakeholders. Deliberation is also viewed as a means to empower citizens by increasing their sense of worth and strengthening their identification with their community (DeLeon, 1990; Dryzek, 1990; Stone, 2002). In the area of STS, public participation is seen as a means to reshape the relationship between experts and the public, and therefore make usually inaccessible but profoundly important science policies and practices more subject to popular direction (Wynne, 1996; Schot and Rip, 1997; Guston and Sarewitz, 2002; Jasanoff, 2003).

The scholarly discussions about deliberative democracy place a high value on 'inclusion' as a measure of good deliberation. The essence of deliberation rests on private individuals coming together as equals to debate and deliberate upon issues of common good. In this formulation of deliberation, status distinctions do not enter the arena to distort deliberation. But many scholars in the field have emphasized that existing social divisions like those based on race and gender can impede the deliberative process (Mansbridge, 1983; Young, 1996; Sanders, 1997). Equality of resources and equal opportunities to participate and weigh different arguments and viewpoints should accompany the development of inclusive deliberation processes (Sanders, 1997).[1]

A variety of criteria have been used by scholars of deliberation to evaluate the levels of inclusion and other qualities of deliberative exercises. Rowe and Frewer (2000), for example, have emphasized that such exercises should meet certain conditions: the participants should be broadly representative of the larger population and they should be involved early in the process; the event should be conducted in an unbiased, transparent and cost-effective manner. Renn et al (1995) have proposed 'fairness' and 'competence'[2] as the normative criteria to be used in the procedural evaluation of deliberation exercises. In addition to these procedural criteria, Burgess and Clark (2006) have proposed productiveness and mutual learning (the extent to which an interaction results in real change and learning on the part of all parties concerned) as criteria for evaluating the outcomes of deliberation.

WWViews is one of the few deliberative exercises in the area of S&T that have taken place in India, and the first that was part of a global initiative. An analysis of the exercise is important to understand its impact as well as how such deliberative processes fit within the larger policy arena in India. The data for my analysis is drawn from three sources: an exit survey that was administered in New Delhi after the event,[3] interviews[4] with the project manager at New Delhi and the chief facilitator at Bangalore, and personal observations at the New Delhi site.

I will analyse the WWViews processes in Bangalore and New Delhi by focusing on the following procedural aspects: differences in the partner organizations and their recruitment methods, group composition of the WWViews panels, roles of facilitation and access to information. I will evaluate the levels of inclusion, fairness and competence of the Indian WWViews processes. I will also explore the (early) learning, media and policy impacts of the Indian WWViews events. In the last section I discuss the strengths and challenges of the Indian WWViews process, and ways towards more effective deliberation models that take into account the Indian (S&T) context of policy-making.

Features of the Indian WWViews process

Unlike many countries that were part of the WWViews project, India does not have much experience with participatory technology assessment processes such as consensus conferences, citizens' juries and scenario workshops. Among the earliest deliberative exercises held in India was a citizens' jury on using GMO crops in small-scale farming. It was organized in a small village in the state of Karnataka in 2000 by the non-profit Actionaid India.[5] This was followed by another deliberative exercise entitled 'Prajateerpu' held in 2001 in the state of Andhra Pradesh on food and farming futures[6] that combined elements from participatory techniques such as citizens' jury and scenario workshops. It was jointly organized by the UK-based International Institute for Environment and Development (IIED), the Institute of Development Studies (IDS) at the University of Sussex, Andhra Pradesh Coalition in Defence of Diversity, the University of Hyderabad and the all-India National Biodiversity Strategy and Action Plan (NBSAP). The citizen jury was made up of small and marginal farmers, food processors and consumers, and discussed three different scenarios regarding food security and agriculture. The results of both these jury deliberations reveal that the participants, many of whom were illiterate or with basic education levels, were able to meaningfully deliberate on technical issues, question experts and put forth a common vision of their future based on their diverse life experiences (Pimbert and Wakeford, 2003a and b). The Indian partners in the WWViews project were a different set of organizations that did not have any prior experience in the area of participatory technology assessment.

Partner organizations

In order to implement the project, the Danish Board of Technology (DBT) partnered with organizations in the participating countries that ranged from universities and civil society organizations to participatory technology assessment institutes. According to DBT, each WWViews partner organization '… should preferably have some experience with citizen participation methods, be unbiased with regards to climate change, be able to follow the common guidelines, and self-finance their participation in WWViews'.[7]

In India, the two partners were the Centre for Studies in Science Policy (CSSP) at Jawaharlal Nehru University, New Delhi, and the Centre for Social Markets (CSM) in Bangalore. CSSP is an interdisciplinary teaching and research centre whose main focus is the relationship between science, technology and society.[8] CSM is a non-profit organization, based in Bangalore, and engaged in the area of climate change. Its Climate Challenge India programme seeks to promote Indian leadership on climate change through 'grassroots work, civic dialogue and business engagement'.[9] Comparing the two organizers, CSSP was a university research centre and somewhat disengaged from the climate change debate, whereas CSM was a non-profit organization actively engaged in the area. These differences were reflected in several ways in how the two organizations arranged the WWViews event. Most importantly, CSSP tried to follow the guidelines established by DBT, viewing it mostly as a research project, while CSM adopted a more flexible approach, and was focused on the opportunity that the event provided for increasing public awareness about climate change and creating 'climate change literacy' among the local populace. It is also important to point out that the two sites did not work in coordination. In fact, there was no communication between the two sites at either the planning or the implementation stages.[10] I was affiliated with CSSP, where I trained facilitators, worked with the event organizers as the only person at the site who had attended the DBT project manager training, and provided support in addition to being an observer on WWViews day.

Methods of recruitment

With regard to the recruitment of participants, DBT had suggested a number of strategies such as placing newspaper advertisements or randomly selecting names from a national register, and constructing balanced panels of 100 citizens with the help of given socio-demographic criteria. The two sites in India adopted very different recruitment methods.

In New Delhi, the organizers decided to contact some registered social organizations, asking them to provide names of their members who might be interested in participating in the WWViews event. This was also considered a way to reach out to citizens living in rural areas and small towns. It was made clear to these organizations that the individuals who were nominated by them would be presenting their individual opinions rather than the viewpoints of the organizations. From the pool of participants nominated by these organizations, invitations to participate were sent to a group of individuals who were diverse in terms of gender, age, occupation and region. Initially the organizers considered placing newspaper advertisements, but finally rejected that option, since they felt that they would be overwhelmed by a high number of applications.

This method of recruitment is similar to the one utilized to recruit participants in 'Prajateerpu', the citizens' jury held in Andhra Pradesh on food and farming futures. Community organizations and local groups were contacted to identify active members who were then interviewed by the researchers to select jurors fitting their criteria – small or marginal farmers who were articulate and lacked close connections to both political parties and NGOs. In addition, emphasis was laid on recruiting women farmers and *Dalit*[11] ['untouchable'] and *Adivasi*[12] ['indigenous'] farmers. The organizers also felt that membership in community organizations was important in terms of the impact of the exercise as participants would be able to utilize membership and network to influence the wider policy debate (Pimbert and Wakeford, 2003a).

In Bangalore, CSM decided to recruit participants by way of snowball sampling, a method of recruiting participants by word-of-mouth through existing acquaintances. They also used the internet to advertise the event by posting information about the event to different email lists. In addition, CSM recruited their corporate partners to publicize the event. To control the recruitment process, the organizers solicited information about the participants with regard to their age, gender and occupation in order to get a diversity of participants.

Group composition

In both New Delhi and Bangalore, the total number of participants was much less than the desired number of 100, which limited prospects for satisfying the already challenging goal of demographic representation. In New Delhi, due to restricted funding, it was decided to aim for a panel of 75 participants. This number of confirmations was achieved, but on the day of the deliberations only 47 participants attended the meeting, a majority of whom had travelled from states other than Delhi. In Bangalore, the number of participants was even less. Only 30 participants attended the meeting. The organizers had sent out invitations to approximately 80 individuals and received confirmations from 40.

The different recruitment methods yielded distinctive types of biases in the structure of the panels at the two sites. In New Delhi, where recruitment was based on contacts to registered social organizations, there was an overwhelming over-representation by men (only 11 per cent of participants were women). Another bias was that the majority of the participants (51 per cent) were under the age of 30. In terms of educational, occupational and regional backgrounds of the participants, however, there was considerable diversity.[13] In Bangalore, where recruitment was based on snowball sampling, better balance between men and women was achieved, but women still represented only one third of the participants. The age of the participants ranged from 18 to 62 years and was skewed towards the 25–35 age group.[14] The low female representation and weighting towards younger populations in both sites can partly be explained by the low turnout rates, which are largely outside the control of organizers, as well as the younger generation's higher involvement in climate change debates in India. A major reason for the low turnout was its timing during the major Hindu festival of Navrartri, which is celebrated for nine nights and ten days across the country during September–October, with the dates varying each year in accordance with the lunar calendar.

The language of deliberation was an important planning issue at both sites, as it is a major contextual factor in multilingual societies that can drastically circumscribe the deliberative space. The New Delhi organizers decided to limit participation to people from the Hindi-speaking states of India due to the difficulty of holding such an event in a multitude of regional languages. Thus, the languages used were English and Hindi in New Delhi and only English in Bangalore. The latter choice effectively excluded the poorer segments of the population from participating at the Bangalore site. According to the organizers, however, within the English panel there was still considerable diversity in terms of age and occupational backgrounds.[15]

In a multilingual society like India, the dominant language of the public sphere privileges its speakers and marginalizes those who do not speak it. As argued by Chandoke (2005, p346), speakers of a subaltern language 'cannot even challenge the conceptualization of the dominant language – not only because their own language may express incommensurate ideas of power and negotiation, but because their language possesses no authority in the domain'. Language not only circumscribes the deliberative space, it is also a source of power affecting the inclusivity of deliberative spaces.

Role of facilitation in deliberation

DBT guidelines called for participants to be seated in groups of six to eight participants with a professional facilitator (or moderator) at each table. The role of the facilitator is usually, as also in DBT guidelines, defined in terms of impartiality. The facilitator has to guide the group to reach a decision without influencing the opinions of the participants. They are needed to ensure that each participant has sufficient time to speak, that there are no unwarranted interruptions, and that particular individuals do not dominate the discourse. In the research literature, facilitation is considered an important mechanism, contributing to the inclusiveness and equality of the face-to-face deliberation process (Hackman, 1987; Offner et al, 1996; Kramer et al, 2001; Creighton, 2005).

In New Delhi, the facilitators were recruited among graduate students. Training sessions were held for the student facilitators to familiarize them with the process and the guidelines for facilitation. A majority was bilingual, which simplified assignments to tables that conversed in Hindi or English only or to tables that were bilingual. In Bangalore, there were no table facilitators, and it was the responsibility of the chief facilitator to moderate the whole discussion process.

The feedback from participants at the New Delhi site was quite positive about the role of facilitation. The organizers at Bangalore also felt that the discussions went well, including the self-moderated groups. They reported that achieving consensus during the recommendation session was easy, and 'the discussions were good-humoured and good-natured'. In the absence of facilitators at the tables in Bangalore, however, a particular challenge was that the groups had to deal with issues of communication bottlenecks and equality of participation by themselves and without impartial external help. With limited data and no comparable control group, however, we cannot decisively conclude that the absence of table facilitators at Bangalore contributed to inequality among the participants or decreased performance.

Access to information

Equal access to information is an important factor contributing to the high quality of deliberation. Access to information is part of the evaluative criterion of competence which refers to the process of verification of knowledge claims (Renn, Webler and Weidemann, 1998; Webler and Tuler, 2000). Availability of background materials is among the most important factors contributing to an informed public opinion as they provide citizens with information that can be utilized while debating the validity of competing knowledge claims.

The two Indian sites differed in terms of how they circulated the background materials to the participating citizens. In New Delhi, participants were provided with the background materials only a day or two before the event. The information materials were translated into Hindi and the videos also had Hindi subtitles. In Bangalore, the participants did not receive the briefing materials at all. To brief the citizens, the chief facilitator made an initial presentation on the topic and the upcoming COP15 meeting in Copenhagen and the videos provided by DBT were used as an additional source of information. The delivery of background information that was either limited (New Delhi) or lacking (Bangalore) meant that the participants had to rely primarily on their previous knowledge about climate change, which caused structural inequality in access to information to be more significant in the proceedings than would otherwise have been the case. The differences in background knowledge of participants between the two sites is clearly indicated by their responses to the WWViews question, 'To what extent were you familiar with climate change and its consequences before joining WWViews?' Of the participants at New Delhi 33 per cent, compared to just 13 per cent at Bangalore, responded that they knew a little or nothing.[16] Early access to the background materials for this segment of the participants would clearly have enhanced their prospects of participating more effectively.

Tracking the impacts

The impacts of deliberative processes can be tracked both in participant learning and in media and policy responses. Participant learning takes place in the context of the deliberation process and can be measured through participant surveys and interviews. Media and policy impacts, on the other hand, take place in multiple contexts of publicity and policy-making, and need to be evaluated over a longer time. In this section I will report on some key findings concerning participant learning based on an exit survey and characterize early media and policy impacts of Indian WWViews based on interviews with national organizers.

Participant learning

Learning is an important aspect of any deliberative exercise. The design of most deliberative exercises rests on the premise that dialogue among a diverse group of participants will lead to the articulation of a variety of values and perspectives. Participants learn not only from information materials but also from the perspectives and experiences of others. Dialogue permits a broader understanding of different interests, leading to learning and a consensus-based decision. Through the process of deliberation, participants learn about other points of view as well as the value of civic participation in general (Burkhalter et al, 2002; Delli Carpini et al, 2004). In New Delhi, 86 per cent of the participants felt that participation enabled them to learn about alternative perspectives to their own, while 82 per cent felt that participation significantly influenced their opinion about climate change. Since it was one of the few deliberative processes in S&T conducted in India, learning about participatory decision-making was as important as learning about climate change. Of the participants, 93 per cent agreed that the exercise helped them learn significantly more about participatory decision-making.

Not only were alternative perspectives heard but they were also valued. Good deliberation requires not only listening to differing viewpoints with respect, but also hearing and debating them. These aspects of learning and reciprocity are what define an inclusive dialogue and are elements that allow public reasoning to be a mode of social justice. A vast majority (86 per cent) of the participants agreed with the statement that participation 'enhanced my understanding of alternative perspectives to my personal opinion on climate change'.

Overall, the evaluations by participants in New Delhi were positive, and 86 per cent of participants stated that the experience of WWViews made them want to participate in another deliberative exercise. The positive estimations by the citizens were echoed in the comments by the interviewed organizers, who felt that the event was successful in allowing participants to express their views and learn from others: 'common people were able to voice their feelings and they were proud to be part of the process' (New Delhi); 'they learned a lot of new things, it gave different perspectives' (Bangalore). Davies et al (2009) point out that even deliberative or dialogical events not linked to the policy process, such as science cafes, are important 'sites for symmetrical individual or small-scale learning'. The social interaction among a diverse group of participants taking place in the context of dialogue and deliberation leads to social learning. The authors, however, warn that this learning can only enhance individual capacity when it is symmetrical in nature and characterized by equality in exchange. The survey data does provide the participants' perspective regarding the matter of learning; however, follow-up interviews with participants after the event and over time could provide a better understanding, and should become a priority in future deliberations.

Media and policy impacts

Despite efforts at both Indian sites, media coverage was non-existent. In New Delhi, the university press office sent out press releases but these were not picked up by the media. The press office also helped to mail the results and recommendations to various policy-makers. Despite lacking media interest at the local and national levels, however, the New Delhi deliberations were covered by a Japanese media team whose counterparts covered sites in Denmark, Japan, Maldives and the US for the production of a documentary that was screened on Japanese public television. The presence of the Japanese media team throughout the day made the Indian participants aware of the fact that WWViews in New Delhi was not an isolated event, but a part of a global project contributing to a global policy process. In Bangalore, invitations were sent out to the press for the deliberation and a press release was also circulated at the conclusion of the event. Reporters from two local newspapers were present, but neither paper published a report on the WWViews event. The organizers also used the internet to disseminate the results to various email lists.

In terms of policy impacts, none of the Indian organizers personally communicated the results to national policy-makers or the country delegates to COP15. One reason is that policy-makers, political representatives and bureaucrats are not easily accessed in India. Setting up appointments requires effort and is often a function of networks, caste and political affiliations, social and economic status, and other similar factors. It also requires experience in negotiating the system. At New Delhi, in particular, the organizers were volunteers taking time out from their academic duties and did not have the time for policy outreach. The latter requires advocacy on the part of the organizers and the advocacy and academician roles often clash. Dissemination of research results through conference presentations and academic writing is different from dissemination of project recommendations to newspapers and policy-makers. Though the task of dissemination was delegated to the university press office with the requisite specialization, it still required advocates of the process, especially since the WWViews process was not just novel to policy-makers but also disconnected from any formal policy process. At Bangalore, the goal of the organizers was primarily to create 'climate change literacy',[17] and dissemination of results to policy-makers was secondary. There it was considered more important to reach out to those who were concerned about climate change and educate them about the issues involved.

In summary, in terms of direct impacts of the WWViews event, measured by the media coverage and presentation of the results to policy-makers, at both Indian sites there was nil impact. It is also important to remember that financial resources were limited at the sites and in terms of prioritization of activities, dissemination was not as high as organizing the event. The WWViews project had a clear goal, which was to present the results to 'politicians, negotiators and interest groups engaged in the UN climate negotiations leading up to COP15 and beyond'[18] so that they are aware of the views of the public. On that account, WWViews in India failed.

Discussion

The challenges in organizing the WWViews deliberations and the limited impact of the event in India can be understood in the broader context and culture of Indian society, and its tradition and style of (S&T) policy-making.

The first point to consider is the huge ethnic and cultural diversity of the country and the increasing inequality in the distribution of resources, both economic and social. India's economic reforms that started in the early 1990s have led to economic growth that has improved the lives of millions of people, but they also contribute to inequality. Over a quarter of the population still lives below the poverty line and the income share held by the lowest 20 per cent of the population was only 8.1 per cent in 2005.[19] The national literacy rate is just 66 per cent.[20] Caste distinctions play an important role in the political arena and difference along the lines of caste can also be identified in political participation. In addition, there is also increasing regional disparity, and the gender gap exists in political participation, both in terms of voting and candidature for elections.

Considering the high levels of diversity and inequity in the Indian society, it is relevant to ask how effectively the method of constructing WWViews type of 'mini-publics' (Goodin and Dryzek, 2006) can represent the populations of a country with a very diverse population structure. Is there any chance of creating an inclusive deliberation process that would – in any reasonable way – represent the voice of the Indian populace?

WWViews was a pilot in global deliberation and therefore the two sites selected in India were not intended to be a final answer to the level and extent to which the voice of the Indian populations should be represented. In any case, it seems evident that in order to include the diversity of perspectives and people that make up India, many more sites should participate in any future global consultation project. Geographically distributing the deliberation processes in India could also address the difficulties experienced in New Delhi in attracting participants from substantial distances outside the urban centre.

Another point of reflection relates to how inequalities in multiple ways are rooted in Indian society, and how they can be compensated for inside and outside deliberative practice. A major cause for concern is the lack of women in the group of participants in New Delhi, despite the concerted attempts of the organizers to select a representative group of participants. Gender inequality is an ever-present reality in India, particularly obvious in the policy-making arena. The absence of women from a deliberative exercise dealing with the topic of climate change is even more of a cause for concern as women, particularly in rural areas, bear the burden of ecological degradation (Agarwal, 1992).

The issue of (in)equality was also approached in the exit survey. A major surprise was that a strong majority of the participants (76 per cent) agreed with the statement: 'important societal groups (ethnic minorities, age and income groups, etc.) were appropriately represented', even though only 11 per cent of the participants in New Delhi were women. One hypothesis is that the presence of a large number of women graduate students in the room, who volunteered as moderators, misled the citizens in their estimations of the actual gender balance. A more frightening possibility to contemplate is that the lack of women in any decision-making forum is taken as a given and does not raise a red flag.[21]

Sporadic deliberation processes cannot provide straightforward means to balance deeply rooted inequalities. Mohanty (2007) has shown that even when women are included in a decision-making process, 'empty spaces' emerge, where they are reluctant to speak and are rarely heard. In her analysis of a community forestry programme in India, Gupte (2004) similarly found that gender stratification resulted in the marginalization of women within the participatory process. In part this reflects women's exclusion from any role in natural resource management decisions even though they have traditionally been responsible for collecting fodder and firewood (Agarwal, 1992). Gupte recommends using 'facilitating policy tools' such as 'the creation of women's self-help groups, conducting separate meetings for women, ensuring a critical mass that emboldens women to speak, recruiting women into the field staff, and small-scale income generating activities for women' (2004, p379). Some of these 'facilitating policy tools' could be built into the design of future deliberation models (Blue et al, Chapter 8*).*

Another issue while discussing inclusion and equality is the identification of the various demographic categories that are relevant in a particular country context so that the group of participants gathered is indicative of the diversity of the country. It is important to note that caste was not explicitly listed as one of the demographic factors that the organizers at either site were considering in the recruitment process. If we are to have a truly inclusive deliberative space in terms of diversity of the major demographic categories in India, then caste should be among the recruitment criteria. It is not that members of each and every caste group have to be included, but it is important that the voices of the *Dalits* ['untouchables'] and those of the *Adivasis* ['indigenous people'] are heard. These groups exist at the margins of the Indian public sphere and any attempt at widening the scope of the debate has to start with them.

A third point to consider is the impact of policy context and prevailing 'civic epistemologies' on public participation and deliberation exercises. According to Jasanoff (2005, p255), 'civic epistemology refers to the institutionalized practices by which members of a given society test and deploy knowledge claims used as a basis for making collective choices'. Science is an important component of civic epistemology but it is not the only one; other elements include assessment styles, local knowledge, policy analysis and the media, as well as public perceptions of policy issues (Miller, 2005).

At the beginning of this chapter, I noted that India has a long tradition labelled by Sen (2009) as 'government by discussion'. Increasingly, however, the process of legitimizing knowledge, especially in S&T, involves a multiplicity of institutions and increasingly takes place away from the public gaze. Undergirding this development is a public deficit model which assumes that public trust in science does not require justification and assessment in the public domain (Visvanathan, 1997, 1998). Rather than encouraging a firm understanding of the potentials and limits of science, one of the primary objectives of the official science policy is developing a scientific disposition in the people and educating them 'to the marvels that science can bring'. The government's S&T Policy 2003 explicitly states 'public awareness of science and technology' as one of its goals: 'Every effort will be made to convey to the young the excitement in scientific and technological advances and to instill scientific temper in the population at large'.[22]

Policy documents acknowledge that ethical and social consequences of technology and science have to be addressed, but the government has done practically nothing to provide or encourage spaces where the public can engage with science. This mindset has spilled over to other policy realms, especially that of technology-based economic development, thus eclipsing India's participatory traditions at the national level. An official discourse that has consistently talked of instilling a scientific temper in its citizens is quite naturally a discourse that also excludes the public from any kind of scientific and technological decision-making. As a result, public perceptions of governance in India reveal that citizens view the government as closed to public input into policy-making (Hyden et al, 2004). Within the area of environmental policy, the government has initiated joint management of forests and watersheds with local communities, but their success has been uneven and often the state has been a major obstacle to realizing goals (Williams and Mawdsley, 2006).

In order for deliberative exercises to achieve the aim of influencing the policy process, there has to be a willingness on the part of bureaucratic agencies to hear what the citizens wish to say. Cornwall and Gaventa (2001, p34) propose that individual 'champions' can have a critical role in promoting change:

> *Where the use of participatory methods for consultation has often been most effective is where institutional willingness to respond is championed by high-level advocates within organizations. Where such 'champions' exist and where they can create sufficient momentum within organizations, the processes of invited participation that they help instigate can make a real difference.*

In their evaluation of the Prajateerpu deliberative exercise, Pimbert and Wakeford (2003a, p40) raise a similar point. They state that 'intermediary individuals and channels' are required to bridge the citizens' jury and those with the power to change policy so that deliberative events do not remain isolated occurrences without impact. It also requires sustained advocacy on the part of these bridging individuals and organizations.

Not only scientific organizations but also the bureaucracy has been impervious to any reflexivity or opening up to viewpoints of the ordinary citizens. There have been some recent attempts at public consultations initiated by the Minister of Environment and Forests on Bt brinjal (genetically modified aubergine) and by the Indian Council of Medical Research and the Department of Biotechnology on guidelines for stem cell research and therapy. It will be interesting to see how these evolve, but until now these have been more in the form of meetings to gather public input on specific policies. These cannot be taken as indications that science policy is moving towards a more democratic approach. The policy framework is so dominated by the view that science is right and scientific progress is certain, that there is no place for reflexive practice to allow assumptions to be re-examined and to deal with the uncertainty of science. The acknowledgement that science in an era of risk and uncertainty requires a diversity of perspectives and that localized knowledge is relevant is a perspective that exists, but it is marginal to the official science discourse. A vibrant tradition reflected in the writings of scholars such as Nandy (1987, 1988) and Visvanathan (1997, 1998), who are critical of this culture of science and system of policy-making, does exist in India. The issue of public accountability is central to their critique, which raises questions about the access and control of knowledge and its connection to gender, class and race; the relationship of emerging technological systems to values of justice, equity and human rights; and the implications of emerging technologies for participatory democracy (Rajan, 2005). In practice, however, India's culture of science policy-making has overshadowed these contemporary critiques.

What is required is a change in the culture of governance, a shift towards what Jasanoff (2003) terms as 'technologies of humility', methods that call for a new relationship between experts, citizens and policy-makers: 'They require not only the formal mechanisms of participation but also an intellectual environment in which citizens are encouraged to bring their knowledge and skills to bear on the resolution of common problems' (Jasanoff, 2003 p227). Design of participatory processes can be adapted to deal with the inequalities and diversity that exist in Indian society, but until the culture of governance changes to view the public as legitimate stakeholders in debates around the design and use of technology, these deliberative exercises can never fulfil their promise of reflexivity and learning.

Deliberative processes such as WWViews can contribute to a change in this culture, but only if they are more than isolated one-off events. Not only is more deliberation needed, but deliberation also requires advocates who will identify and partner with those inside the government who are both willing and able to help promote the cause of citizen participation (the 'champions' cited by Cornwall and Gaventa, 2001). These advocates should also link up with contemporary Indian critiques of the dominant policy culture. This will not only increase citizen participation in science policy formulation, it will also advance the quintessentially Indian tradition of argumentation and dissent.

The gap between participatory and formal democracy that exists in India can only be reduced by 'opening up' (Stirling, 2008) the policy process to a wider range of voices and perspectives. This is particularly important as globalization and economic reforms are rapidly creating an India in which large sections of the population are marginal to the prosperity that rapid economic growth is bringing. Science and technology are important drivers of this growth, creating a society that is certainly not reflective of the vision of a large proportion of the population. Participatory technology assessment, which is ultimately about the type of society citizens want, becomes even more imperative in the rapidly globalizing India. Social justice requires that public deliberation expands not just in terms of topics but also of participants.

The experience of WWViews India shows that organizers' decisions, a standardized format, limited financial resources, the wider policy context and existing social
inequities all hampered the inclusivity of the process. The design of these participative
exercises must be carefully thought out so that they are inclusionary reflexive spaces allowing for dissent and dialogue, not existing merely to create justification for particular
policies or to feign adherence to the ideal of deliberative democracy.

Notes

1 Verba et al (1995) have pointed out that education is a major contributor to inequalities in
   deliberation. The well-educated have better reasoning skills, which has an impact on the
   argumentative aspect of public deliberations, whereas less-educated persons do not have
   access to occupations where reasoning skills develop and also lack access to the available
   public arguments around these issues (Nie et al, 1996; Mendelberg, 2002).
2 Fairness refers to equal opportunities for all participants to attend the exercise and initiate discussion, as well as to debate the knowledge claims. Competence is defined as 'the
   construction of the best possible understandings and agreements given what is reasonably
   knowable to the participants' (Webler, 1995, p65).
3 A similar exit survey was administered at 21 sites in 16 countries across the globe.
4 The interviews were carried out by Dawn Bickett, another WWViews researcher, and me.
   Neither interviews with participants nor exit survey results from Bangalore were, unfortunately, included in this study.
5 http://actionaid.org.uk/doc_lib/citizens_jury_initiative.pdf.
6 Project details are available at www.prajateerpu.org.
7 www.wwviews.org/node/249.
8 The Jawaharlal Nehru University, with which CSSP is associated, is one of India's premier
   universities, drawing students from all parts of India, and is one of the most diverse campuses
   in India in terms of student profiles.
9 In December 2007 Climate Challenge India was profiled at the UN Climate Summit in Bali,
   Indonesia and in December 2009, the focus of the initiative was to build the conditions necessary for a successful outcome to the crucial United Nations negotiations on climate change
   (COP15) in Copenhagen.
10 This is in sharp contrast with the US, another country with multiple sites, where the five
   WWViews sites were in contact with each other from the initial planning stages to later
   stages of evaluation and dissemination.
11 *Dalit* is a term for the numerous caste groups that are considered as 'untouchables' in the
   traditional caste system.
12 *Adivasi* is a term used for a heterogeneous group of ethnic and tribal groups who are considered to be the original inhabitants of India.
13 Based on an interview with a project manager at New Delhi.
14 The source of the demographic data for the New Delhi participants was the exit survey
   completed by the participants. In the case of Bangalore I have used information provided by
   the organizers.
15 Based on interviews with a Bangalorean facilitator and an organizer.

16 Responses to the questions are available at the WWViews website. www.teknov2.tdchweb. dk/new2/index.php?cid=1273&gid=blank&ccid=1272&cgid=blank&question=blank&rec= 0&lang=573&reclang=0.
17 Interview with Bangalore site chief facilitator.
18 WWViews project, www.wwviews.org/node/194.
19 World Bank's World Development Indicators Database, September 2009; available at http:// databank.worldbank.org/ddp/home.do.
20 United Nations Human Development Index, available at http://hdr.undp.org/en.
21 The question is focused on representation of 'important social groups' but does not list women as an example. A higher percentage of respondents might have noted their absence had there been a cue about gender in the question, but the failure to observe the obvious gender inequality without a cue further strengthens the claim that this inequality is taken for granted and would only be observed (if at all) when explicitly brought to the respondent's attention.
22 Government of India's Science and Technology Policy 2003, available at www.dst.gov.in/ stsysindia/stp2003.htm.

References

Agarwal, B. (1992) 'The gender and environment debate: Lessons from India', *Feminist Studies*, vol 18, no 1, pp119–156

Bhargava, R. (2005) 'Introduction', in R. Bhargava and H. Reifeld (eds) *Civil Society, Public Sphere and Citizenship: Dialogues and Perceptions*, Sage, New Delhi

Bhargava, R. and Reifeld, H. (eds) (2005) *Civil Society, Public Sphere and Citizenship: Dialogues and Perceptions*, Sage, New Delhi

Burgess, J. and Clark, J. (2006) 'Evaluating public and stakeholder engagement strategies in environmental governance', in A. G. Pereira, S. Vaz and S. Tognetti (eds) *Interfaces between Science and Society*, Greenleaf Press, Sheffield

Burkhalter, S., Gastil, J. and Kelshaw, T. (2002) 'A conceptual definition and theoretical model of public deliberation in small face-to-face groups', *Communication Theory*, vol 12, pp398–422

Chandhoke, N. (2005) 'Exploring the mythology of the public sphere', in R. Bhargava and H. Reifeld (eds) *Civil Society, Public Sphere and Citizenship: Dialogues and Perceptions*, Sage, New Delhi

Cornwall, A. and Gaventa, J. (2001) 'Bridging the gap: Citizenship, participation and accountability', *PLA Notes*, vol 40, pp32–35

Creighton, J. L. (2005) *The Public Participation Handbook: Making Better Decisions Through Citizen Involvement*, John Wiley & Sons, San Francisco

Davies, S., McCallie, E., Simmonson, E., Lehr, J. and Duensing, S. (2009) 'Discussing dialogue: Perspectives on the value of science dialogue events that do not inform policy', *Public Understanding of Science*, vol 18, no 3, pp338–353

DeLeon, P. (1990) 'Participatory policy analysis: Prescriptions and precautions', *Asian Journal of Public Administration*, vol 12, no 1, pp29–54

DeLeon, P. (1997) *Democracy and the Policy Sciences*, State University of New York Press, Albany, NY

Delli Carpini, M. X., Cook, F. L. and Jacobs, L. R. (2004) 'Public deliberation, discursive participation, and citizen engagement: A review of the empirical literature', *Annual Review of Political Science*, vol 7, pp315–344

Dryzek, J. S. (1990) *Discursive Democracy: Politics, Policy, and Political Science*, Cambridge University Press, Cambridge, UK

Dryzek, J. S. and Tucker, A. (2008) 'Deliberative innovation to different effect: Consensus conferences in Denmark, France, and the United States', *Public Administration Review*, vol 68, no 5, pp864–876

Einsiedel, F., Jelsøe, E. and Breck, T. (2001) 'Publics at the technology table: The consensus conference in Denmark, Canada, and Australia', *Public Understanding of Science*, vol 10, no 1, pp83–98

Fischer, F. (2000) *Citizens, Experts, and the Environment: The Politics of Local Knowledge*, Duke University Press, Durham, NC

Goodin, R. E. and Dryzek, J. S. (2006) 'Deliberative impacts: The macro-political uptake of mini-publics', *Politics and Society*, vol 34, no 2, pp219–244

Gupte, M. (2004) 'Participation in a gendered environment: The case of community forestry in India', *Human Ecology*, vol 32, no 3, pp365–382

Guston, D. H. and Sarewitz, D. (2002) 'Real-time technology assessment', *Technology in Society*, vol 24, no 1–2, pp93–109

Hackman, J. R. (1987) 'The design of work teams', in J. W. Lorsh (ed) *Handbook of Organizational Behavior*, Prentice Hall, New York

Hyden, G., Court, J. and Mease, K. (2004) *Making Sense of Governance: Empirical Evidence from Sixteen Developing Countries*, Lynne Rienner Publishers, Boulder, CO

Jasanoff, S. (2003) 'Technologies of humility: Citizen participation in governing science', *Minerva*, vol 41, no 3, pp223–244

Jasanoff, S. (2005) *Designs on Nature: Science and Democracy in Europe and the United States*, Princeton University Press, Princeton, NJ

Kramer, T. J., Fleming, G. P. and Mannis, S. M. (2001) 'Improving face-to-face brainstorming through modeling and facilitation', *Small Group Research*, vol 32, no 5, pp533–557

Mansbridge J. (1983) *Beyond Adversary Democracy*, University of Chicago Press, Chicago

Mendelberg, T. (2002) 'The deliberative citizen: Theory and evidence', in M. X. Delli Carpini, L. Huddy and R. Shapiro (eds) *Political Decision-Making, Deliberation and Participation*, JAI Press, Greenwich, CT, pp151–193

Miller, C. A. (2005) 'New civic epistemologies of quantification: Making sense of indicators of local and global sustainability', *Science, Technology & Human Values*, vol 30, no 3, pp403–432

Mohanty, R. (2007) 'Gendered subjects, the state and participatory spaces: The politics of domesticating participation in rural India', in A. Cornwall and V. S. Coelho (eds) *Spaces for Change? The Politics of Citizen Participation in New Democratic Arenas*, Zed Books, London

Nandy, A. (1987) *Traditions, Tyranny, and Utopias: Essays in the Politics of Awareness*, Oxford University Press, New York

Nandy, A. (1988) *Science, Hegemony and Violence*, Oxford University Press, Delhi

Nie, N. H., Junn, J. and Stehlik-Barry, K. (1996) *Education and Democratic Citizenship in America*, University of Chicago Press, Chicago

Nielsen, A. P., Lassen, J. and Sandoe. P. (2007) 'Democracy at its best? The consensus conference in a cross-national perspective', *Journal of Agricultural and Environmental Ethics*, vol
20, no 1, pp13–35

Offner, A. K., Kramer, T. J. and Winter, J. P. (1996) 'The effects of facilitation, recording and
pauses upon group brainstorming', *Small Group Research*, vol 27, no 5, pp283–298

Pimbert, M. and Wakeford, T. (2003a) 'Prajateerpu, power and knowledge: The politics of
participatory action research in politics; Part 1, Context, process and safeguards', *Action
Research*, vol 1, no 2, pp184–207

Pimbert, M. and Wakeford, T. (2003b) 'Prajateerpu, power and knowledge: The politics of
participatory action research in politics; Part 2, Analysis, reflections and implications', *Action
Research*, vol 2, no 1, pp25–46

Rajan, S. R. (2005) 'Science, state and violence: An Indian critique reconsidered', *Science as
Culture*, vol 14, no 3, pp265–281

Renn, O., Webler, T. and Wiedemann, P. (eds) (1995) *Fairness and Competence in Citizen
Participation: Evaluating Models for Environmental Discourse*, Kluwer Academic, Boston

Rowe, G. and Frewer, L. J. (2000) 'Public participation methods: A framework for evaluation',
Science, Technology & Human Values, vol 25, no 1, pp3–29

Sanders, L. M. (1997) 'Against deliberation', *Political Theory*, vol 25, no 3, pp347–376

Schot, J. and Rip, A. (1997) 'The past and future of constructive technology assessment',
Technological Forecasting and Social Change, vol 54, no 2, pp251–269

Sen, A. (2005) *The Argumentative Indian: Writings on Indian Culture, History and Identity*,
Penguin Books, New Delhi

Sen, A. (2009) *The Idea of Justice*, Allen Lane, London

Stirling, A. (2008). '"Opening up" and "closing down": Power, participation, and pluralism
in the social appraisal of technology', *Science, Technology & Human Values*, vol 33, no 2,
pp262–294

Stone, D. A. (2002) *Policy Paradox: The Art of Political Decision Making*, Norton, New York

Verba, S., Scholzman, K. L. and Brady, H. E. (1995) *Voice and Equality: Civic Voluntarism in
American Politics*, Harvard University Press, Cambridge, MA

Visvanathan, S. (1997) *A Carnival for Science: Essays on Science, Technology and Development*,
Oxford University Press, New Delhi

Visvanathan, S. (1998) 'A celebration of difference: Science and democracy in India', *Science*,
vol 280, no 5360, pp42–43

Webler, T. (1995) '"Right" discourse in citizen participation: An evaluative yardstick', in
O. Renn, T. Webler, and P. Wiedemann (eds) *Fairness and Competence in Citizen Participation: Evaluating New Models for Environmental Discourse*, Kluwer Academic, Dordrecht

Webler, T. and Tuler, S. (2000) 'Fairness and competence in citizen participation: Theoretical
reflections from a case study', *Administration and Society*, vol 32, no 5, pp566–595

Webler, T., Tuler, S. and Krueger, R. (2001) 'What is a good public participation process? Five
perspectives from the public', *Environmental Management*, vol 27, no 3, pp435–450

Williams, G. and Mawdsley, E. (2006) 'Postcolonial environmental justice: Government and
governance in India', *Geoforum*, vol 37, no 5, pp660–670

Wynne, B. (1996) 'Misunderstood misunderstandings: Social identities and public uptake of
science', in A. Irwin and B. Wynne (eds) *Misunderstanding Science? The Public Recon-
struction of Science and Technology*, Cambridge University Press, Cambridge, UK

Young, I. M. (1996) 'Communication and the other: Beyond deliberative democracy', in S. Benhabib (ed) *Democracy and Difference: Contesting the Boundaries of the Political*, Princeton University Press, Princeton, NJ

10

Linking Citizen Participation and Education in Sciences: The Case of Uruguay

Isabel Bortagaray, Marila Lázaro
and Ana Vasquez Herrera

Uruguay is a small country with fewer than 3.5 million people located in the temperate zone of South America. Several characteristics make Uruguay an interesting site for the WWViews experiment: it has a highly centralized structure, with about half the population living in or around the capital city Montevideo; a strong civilian democratic tradition that includes mandatory suffrage; and a widespread participatory attitude towards political issues. The political character of Uruguayans goes in hand with a tradition of an established welfare state that has anticipated numerous societal demands since its origins in the early 20th century. However, in some policy areas, namely technology, innovation and environmental issues, the state has not had a very active role until lately.

In recent years this situation has changed. In 2009 after a severe drought, the government created the National System of Response to Climate Change and Variability to co-ordinate and plan the required public and private actions for risk prevention, mitigation and adaptation to climate change (Decree of the Executive, 2009). As a consequence, a Plan for Strategic Action was created to analyse the effects of climate change and establish strategies for adaptation in the key areas for our country: the productive sector, health, urban development, energy, biodiversity and ecosystems.

The concern about climate change and environmental issues has also been tackled by civil society through the action and research of non-governmental organizations (NGOs) and the University of the Republic. In 2008 a group of Uruguayan sociologists, biologists, communication studies and science, technology and society (STS) studies teachers, audiovisual producers, graphic designers, cultural managers, photographers, writers and scriptwriters created 'Simurg', a non-profit civil association aimed at facilitating interaction between society and the actors involved in the production and management of technological and scientific knowledge. *Simurg* is the name of a legendary bird possessing knowledge of the ages and therefore symbolizes the organization's mission to enhance the transfer and social appropriation of knowledge. Simurg takes a participatory approach to this mission by engaging different social stakeholder groups, not only in processes of knowledge generation, but also in its communication and dissemination. Currently the organization is working in two main areas: one linked to the expansion of opportunities for citizen participation in the field of science, technology and innovation, and the second related to the development of participatory projects on biodiversity conservation involving children from several schools across the country.

Given this mission, Simurg's members unanimously accepted the Danish Board of Technology's (DBT) invitation to take part in WWViews, which required a range of skills that aligned well with the members' profiles. From the beginning of the project, links were established with other organizations, especially the University of the Republic, Uruguay.[1] The University link allowed the project to be developed within a framework of academic reflection and education that informs the analysis in this chapter, even though this was not the authors' original intent.

The WWViews project in Uruguay therefore created an opportunity to link participatory science and technology (S&T) initiatives with research in this field. In fact, part of the Simurg team has continued the theoretical and practical exercise in their work at the University by organizing a new initiative in public participation: a consensus conference on nuclear energy in Uruguay (Hirschfeld, 2010).

In the following sections of this chapter we analyse the strategy and implementation of the Uruguayan WWViews process, and discuss how the project contributed to social learning and appropriation of science and technology policy issues in the nation. The Biology and Society course is used to illustrate how abstract ideals of science in society dialogue can be connected to practical action.

Simurg's vision of science and public engagement

Simurg's mission is rooted in a particular understanding of scientific activity. In contrast to a traditional conception of science[2] as an apolitical activity, aiming at progressive generation of objective knowledge through universal research methods (e.g. Merton, 1973), Simurg ascribes to a view of science as social process, where non-epistemic elements play an important role in the genesis and consolidation of its products. Following this view, it becomes essential to develop new approaches that facilitate interaction between scientific institutions and their societal constituencies, including ordinary citizens.

The concept of 'social appropriation of science' involves the idea that science is not an exotic undertaking of insulated experts and their elite patrons, but rather activity connected to the broader society through funding and impacts. According to this view, science should be of concern to society at large, and social appropriation of scientific and technological knowledge can further be stimulated by the active participation of non-scientific actors. The concept of social appropriation goes beyond the terms of popularization and public understanding, since it refers to more complex processes than scientific diffusion, including strategies for society to use and take advantage of S&T results and its ability to decide on those issues (Einsiedel and Eastlick, 2000; Wachelder, 2003; González García et al, 2006; Cerezo and Cámara, 2007).

To advance social appropriation of science, it is necessary to rely on a new social contract for science that facilitates the integration between the S&T system and society, one that guarantees access to and valuation of lay knowledge. General education in sciences is an important means to provide the view of science that goes beyond its cognitive content by taking into account its ethical, economic, political, cultural, socio-logical, legal, historical and epistemological aspects and implications.

The approach of Simurg, and the authors of this chapter in particular, is based on this theoretical framework which, unlike more traditional approaches to science, com-plies with the post-normal – or 'science with the people' – epistemological approach (Funtowicz and Ravetz, 2000). In the following sections of this chapter we will analyse how the implementation of WWViews in Uruguay reflected our vision of science and public engagement, and how the WWViews process was integrated with science education.

The strategy and implementation of WWViews Uruguay

The main strategies that guided WWViews Uruguay were:

• a participatory and collaborative approach to all aspects of the project;
• close monitoring of and responsiveness to the local context and its dynamics;
• a sustained focus on the quality of the participants' experiences;
• reflection on the goal of social appropriation as an integral part of the project.

Resources were a significant challenge from the outset. Given its creation just months before the WWViews project launch, Simurg had few of the institutional connections normally required to secure funding. However, the flexibility of its members and their interest in having the public deliberation conducted in the country made it possible to begin the project without initial funding.

All members of Simurg participated actively in the organization of WWViews, and only one external contract was made with a manager in charge of finalizing the details for the event. The project was coordinated by one of Simurg's members while others assumed responsibility for specific functions, such as recruitment, logistics, publicity, venue, etc. This reflected a conscious strategy to achieve a level of engagement and commitment that would lead all those involved to feel part of the process, thereby enhancing collective learning. The same strategy was successfully followed for the stakeholders who gradually became involved in the project as well.

In keeping with the collaborative emphasis, financial and institutional support were treated as interdependent issues, with institutional support possibly leading to financial support. A political environment that was newly receptive to S&T and a government that was easily accessible provided an advantageous context for organizing a series of meetings with the Mayor of Montevideo, the Minister of Environment, the Director of S&T (the head of a public office within the Ministry of Education and Culture) and the University's Rector.[3] Meetings were also held with representatives from the United Nations Development Programme and members of Parliament's S&T committee. Even though no financial support was achieved through these contacts, they generated in-kind support for critical items such as logistics and venue. The latter, for example, was offered by the Parliament, whose explicit support was thus made visible to all by the location of the event. Ultimately our strategy of knocking on many doors yielded financial results in the form of funding from the Danish Embassy in Buenos Aires (which manages relations with Argentina, Uruguay and Paraguay).[4] Overall, this US$16,000 contribution covered all aspects of the 26 September event (except for the venue and the food, which were provided in kind by Parliament and the local government of Montevideo, respectively), plus audiovisual and written dissemination materials.

Collaboration was also critical in the recruitment of citizens. Considering the extremely centralized distribution of the Uruguayan population, recruitment strategy outside Montevideo called for establishing partnerships with existing networks to reach various areas of the country. Especially helpful were Centros MEC, a network of cultural units established in recent years in various localities to improve communication throughout Uruguay (many of them working as social centres, cybercafés, etc.), and the network of managing officers for a national S&T programme.

The focus on a positive experience for everyone connected with the project was sustained from start to finish. To register participating citizens, for example, a blog was created with background information and an online registration link. In light of Uruguay's wide digital gap, however, alternative registration methods were offered by text message and telephone.

Contrary to our expectations, recruitment in the capital city of Montevideo was much more challenging than in the rest of the country. This was in part due to the fact that participation was far less attractive for people living in the capital, as it did not entail the novelty of travelling, staying at a hotel and sightseeing that would be enjoyed by participants from other cities, towns and rural areas. This opportunity for residents from outside the capital, on the other hand, was contingent on the effective management of their travel arrangements, and could readily become a liability for them and the project if accommodation seemed unsanitary or unsafe, or travel arrangements were beset by poor scheduling or long delays. The project team therefore took special care to assure that each participant's experience was positive in this and every aspect of the deliberation.

Close attention and responsiveness to the local context were important in many aspects of the project. For example, Simurg's media strategy recognized that climate change had previously received little attention in Uruguay, and that 26 September happened to be National Heritage Day. The latter is an extremely important and popular day in the country, when institutions open their doors to the public and organize various recreational and cultural activities. Acknowledging that the media would under these conditions be focused on events that are more dramatic or resonant with local traditions than WWViews, we tried to frame the interesting story behind WWViews, not through the topic (climate change), but through the methodology, since it was the first time that a large-scale citizen consultation was attempted in Uruguay. We also invited some prominent speakers, such as the Danish Ambassador to Argentina, a member of Parliament and the Rector of the University, as well as representatives from the local government of Montevideo, to give the opening and closing speeches at the event. Anticipating that media coverage would in any case be sparse, we also decided to budget for the development of dissemination materials that could be used afterwards (a 20-minute audiovisual documentary and a written account with an emphasis on citizen participation in matters of S&T were later produced and distributed to educational, social, political and cultural organizations).

Finally, connecting the participatory and reflective components of Simurg's vision, the views of the participating citizens were integrated into many aspects of the project and the resulting educational materials. For this purpose, an evaluation was carried out in the form of an exit survey[5] where participants could express their own views about the different aspects of the process. The Biology and Society seminar at the University was the centrepiece of our strategy to make social appropriation integral to all aspects of the project, so we address below its organization, and its larger strategic significance later in this chapter.

Biology and Society: Imprinting a science–society dialogue on university education

One of the first steps in organizing the forum in Uruguay was to design a seminar on Biology and Society at the School of Sciences. The course was designed as part of the curriculum of the Biology degree for 12 students for four months. During this period, students would participate in the preparation, development and evaluation of the citizen forum as facilitators. The themes of climate change and global warming would serve as the link to a broader reflection on the role of science and politics in socio-environmental issues.

Every week during the seminar, students discussed the theoretical frameworks within which public participation and communication activities would be set up. This formal process enabled students to get credits for the seminar that would contribute to their degree while also exposing them to practical experiences in the relationship between science and society through participatory mechanisms.

During the first phase of the seminar prior to the event, weekly meetings were devoted to the themes of the forum: climate change as a socio-environmental problem, links between science and politics, and arguments for promoting public participation on S&T. The students proposed research projects on the topic of climate change in Uruguay as a way of embedding the global project in the local context. Students engaged with different problems: some focused on the status of climate change in Uruguay, others did a summary of the WWViews information materials that was sent to the participating citizens beforehand. They worked in groups around their particular interests, and during this first phase other themes relating to the different approaches to science and participation in S&T were discussed. The general methodology remained the same for three months, while students participated in workshops and roundtables aimed at fostering collective work and reflection upon these issues.

As 26 September approached, the role of the facilitator became the focus of the seminar. Students also participated in recruitment of participants and the dissemination of results as part of their practical work within the programme. Theoretical reflection arising from the activities was centred on questioning the way in which lay citizens could contribute to S&T arguments. Do citizens have something to say about complicated scientific questions? Why might they think they do not have the right to participate or decide? What image of science maintains the idea that society and S&T are disconnected? Is it necessary to have previous knowledge about scientific issues to participate?

After the WWViews event, the biology students participated in the evaluation of the process and results of WWViews, both from Uruguay and around the world. They also prepared a poster, on which the results of WWViews consultation were presented to other students at the School of Sciences. Some students also participated in the presentation of results to public authorities and to the Uruguayan delegation to COP15.

Results of the strategies: Implications for social learning

Citizen participation has been promoted using democratic and substantive arguments (Agger et al, Chapter 3, this volume). Promoting participation for social learning and acting on the outcomes is an extension of the substantive argument proposed by Fiorino (1990). This argument recognizes the importance of social learning itself, understood as a process resulting from the behaviour of a group immersed in a common process of learning (Einsiedel and Eastlick, 2000).

From this point of view, social learning is not only an interesting by-product of participatory initiatives but an important end in itself in order to adapt and transform towards sustainability. This is not only due to the pertinence of participants' knowledge but also the benefits of collaborative learning involving all actors (Lázaro-Olaizola, 2009).

We next discuss the aspects of social learning by reviewing the views of the organizers, participants and students about the different aspects mentioned so far, in the light of the theoretical framework that informs this chapter. The views of the participants were taken from the results of the above-mentioned evaluation survey.

According to the evaluation carried out among participants at the end of the event, social learning was perhaps one of the most important results of WWViews Uruguay.

Evaluation results showed an extremely positive view of the forum and the levels of participation that were achieved. All the respondents believed that this type of event should be organized to discuss other topics and would like to participate again. Among the comments there were several messages of gratitude for the opportunity to participate and many recommendations propose collective actions around the topic of climate change (for example, street demonstrations during COP15, establishment of social networks to continue the discussions, an interest group representing the citizens' visions on the topic). These comments reflect the substantial enthusiasm and motivation that the organizers noted throughout the consultation.

There are many aspects which explain this: sharing a collective experience, exchanging ideas and constructing concepts within a team, being part of a bigger initiative with a noble aim. With regards to event organization, it could be argued that the special care and dedication devoted to organizing the forum made participants feel comfortable and created a positive working environment. The organization was highly praised in the evaluation and in fact nothing went wrong in that respect. Transportation arrangements were precise and punctual, and the whole event took into account the needs of the participants. This prevented loss of precious time and energy and ensured that the organizing team was in good spirits to work and enjoy the day.

The recruitment process was also successful, both in the view of the organizers and the participants, as the evaluation results indicate. A total of 96 people turned up and took part. As recommended by the methodology, great care was taken to select a group of people representing a diversity of geographical location, level of education, sociocultural background, age and gender. About half the participants came from the capital city of Montevideo, reflecting how the population is distributed in Uruguay. The rest came from other cities, towns or rural areas widely spread over the country. The number of absentees could not be completely compensated for and the event ran short by four participants, all from Montevideo. Most of the participants (92 per cent) felt there was a wide variety of people at their tables and that a variety of opinions were represented.

Discussions ran smoothly, with all respondents feeling that their own opinions and those of all other people at their tables were heard. As expected by the organizing team, most participants (56.5 per cent) were already aware of most of the information provided on climate change before and during the event. On the other hand, a substantial minority (34.8 per cent) was not. The variety of participants supports the conclusion that the recruitment process was satisfactory. The collective positive feeling and unity that the participants expressed at the end of the day leads us to conclude that the event was successful in achieving good, varied and harmonious exchanges of opinion among a highly diverse group of people.The organizational and recruitment success must have contributed to a certain degree to the effect that the event had on the participants. The results and analysis of the questionnaires returned at the end of the day show that 92 per cent of participants were motivated to learn more about the topic as a result of the activity. Combined with the 28 per cent who mention a change of opinion with respect to climate change as a result of WWViews, this shows that the forum had a great impact on their understanding of science. This supports the aforementioned idea that participatory initiatives have the potential to generate and shape scientific culture, especially among civil society actors.[6]

There were also other aspects in the WWViews process that prompted reflection among the organizers. Notwithstanding the unfortunate overlap with National Heritage Day on 26 September, the sparse media coverage – in spite of considerable effort – was an issue of discussion. The lack of interest by the media in environmental topics was also raised by one of the government representatives as a very difficult problem to overcome. This could not be explained solely by ignorance or indifference on the part of the media, but as part of a wider problem that transcends our local media. In being alienated from direct sources of scientific information, the media tend to report only what is already digested for them, usually whatever happens elsewhere in the world (Haro, 2003). Cultivating friendly relationships with reporters who are familiar with the type of work Simurg is doing was acknowledged as a way of increasing media attention in the future. The dissemination materials planned as part of this project are expected over time to further compensate for the lack of media coverage by reaching at least part of the intended audience: politicians, educators, NGOs and participants. These efforts have already allowed Simurg to maintain contact with the participants while also giving them an idea of the scope of the project beyond the actual consultation.

The fact that funding came from outside Uruguay also raises important questions, particularly with regard to the friendly attitude towards S&T that currently exists in Uruguay and the interest in participatory methodologies that was expressed by all those we engaged during the course of this project. The topic of climate change was only marginally included on political agendas until recently, when a severe drought affected the country, and this prompted the creation of a governmental emergency committee to mitigate and combat climate change, as described above. This fact, together with the innovative mechanism of citizen consultation, raised interest in the project but could not progress beyond this because the expertise and mechanisms to embrace the project and provide it with solid institutional support are not yet in place.

These observations imply that this could be the beginning of an important process. This event has encouraged our interest in pursuing this line of work and in organizing more consultation events on other topics, the next one being nuclear energy. Our aim is to begin shaping our political system around these matters to make it more sensitive to citizens' views on S&T. The audiovisual documentary and written account therefore confirm the status of WWViews Uruguay as a precedent for new initiatives, and also provide important tools for implementing such initiatives by describing the method, procedures and outcomes of the project.

Biology and Society seminar

Integrating the seminar on Biology and Society into the WWViews forum was, in our opinion, one of the most significant achievements of the project in Uruguay. Together with a theoretical reflection upon the links between science, technology and society and the concepts of science and its social contracts, the seminar allowed students to link their reflections with a practical exercise. The students experienced issues surrounding science democratization, the relation between science and society, and the social learning involved in a concrete participatory mechanism. This type of work goes on at the Department of Science and Development's School of Science, offering social programmes within the framework of STS studies for all the science students (Davyt and Lázaro, 2009); but this was the first instance of relating a social approach specifically to a biology course.

The progress shown by the 12 first-year biology students throughout the trimester was striking. The project increased their commitment to science in society issues, and several of the proposals associated with the WWViews project in Uruguay were later initiated by the students. Their keenness to continue their work around these topics (as part of Simurg for instance, or within the activities of community outreach organized by the School of Science) demonstrates the interest they developed in the interactions between science and society.

It is important to mention the views held by the participants about facilitation throughout the day. A large majority (90 per cent) thought that the facilitator at their table did a good job of managing the discussions. Some of these views also appear in the comments section of the questionnaires, where participants specifically congratulate the students.

In addition to a thoughtful analysis of the social contexts of science and technology, the STS approach requires students to reflect creatively on ways to articulate science education, the relationships between society and environment, and the conditions where debate about science, ethics and politics can be established. We believe that the seminar on Biology and Society set the stage for all these possibilities.

Introducing science students to these concepts early in their careers adds an essential dimension to their profiles as future scientists and means that, from the start, these students will consider the bigger picture when undertaking their scientific activities. As a result of their participation in the seminar and the forum, they possess the tools required to incorporate reflection about the concept of the societal role of science as part of their professional practice. As they progress in their careers, the scientific knowledge they absorb from now on will be considered from this new point of view they have acquired, both theoretically and practically, as facilitators of the WWV forum. Through debate and exchange of opinions they are able to convert their fellow students and future colleagues, who have not been exposed to these concepts in the Uruguayan science education system. In their own words:

> *[the forum] ... provided a better place for the students as responsible citizens, giving them the opportunity to change the society in which they are immersed, making them aware that not only the experts have the right to express their opinion.*

Policy implications and the road ahead

The relationship between science and politics is acutely obvious when it comes to environmental issues. Science has played a key role in discovering and publicizing the ozone layer, identifying climate change, and advancing a proposal for preserving biodiversity in the global political agenda. However, more science does not necessarily mean better political decisions. To wait for good science to make decisions could lead to obstacles and delays for relevant issues, such as pollution, acid rain or even climate change (Jasanoff and Martello, 2004).

Particularly when it comes to complex issues of technological and environmental risks that are characterized by high uncertainty, science is not always able to provide unequivocal interpretations and practical recommendations. An additional layer of complexity is involved because environmental laws and regulations do not recognize national or local borders. The 'borderless', long-term and systemic character of the issue, together with the difficulty in establishing a link between human actions and environmental effects makes environmental regulation very challenging, and calls for a profound change in our developmental and behavioural models. Thus, effective environmental governance demands close collaboration between citizens and civil society at all stages of the policy process, from the initial consultations to policy implementation and evaluation.

Several scholars of deliberative governance have focused on the analysis of how participation could be institutionalized (Fuller, 2006). They acknowledge that with institutionalization comes the risk of turning meaningful participation into a rhetorical exercise with no real impact on the social construction of knowledge or decision-making processes (Martin and Sherington, 1997). To avoid this, institutions must embrace innovations and changes that will lead to the stimulation of a true participatory culture. The WWViews project, as the first large exercise of deliberative democracy around the globe, is an example of the relevance and feasibility of these types of participatory mechanisms and their potential to transfer and adapt to new local contexts and cultures. Furthermore, fostering new communication and participatory channels for society could stimulate and raise awareness and curiosity for learning, appropriating and questioning scientific knowledge.

The WWViews consultation was the first exercise of this type ever implemented in Uruguay and has paved the way for other similar exercises, such as the consensus conference on nuclear energy conducted in autumn 2010 by the Science and Development Unit of the School of Science. Its only precedent was a series of public audiences coordinated by the Ministry of Environment, a top-down process, to get stakeholders' voices in a decision-making process regarding the establishment of protected natural areas in the country. It could be said that Uruguay is taking its first steps towards active public participation in S&T matters, thus stimulating the transition of political and academic discourses around participation into concrete action, while introducing the new dimension of S&T into these discourses, which are traditionally centred on other social aspects considered of public interest.

This transition process is already raising many issues, challenging the traditional visions of a science that is separate from its social context. The key question perhaps is: How and what can an ordinary citizen contribute to S&T issues? This question became the central focus of the process and was raised by science students, politicians and journalists alike. It shows how the traditional concepts of the nature of science underlie our culture and are still present, especially in science education, although they are not necessarily explicit or a product of analysis and reflection. Who are these citizens that participate? Why should their vision influence processes under technical and political control? How deep should the discussions go? Can lay people manage technical aspects or should they restrict themselves to the social effects of S&T?

In our opinion these are crucial questions to be analysed and reflected upon in the public arena, particularly as part of the early stages in science education. We feel that the way in which we organized the WWViews process in Uruguay has made progress in that direction.

The WWViews initiative also raised critical evaluation of its methodology: the fact that the citizens are free of other constraints that may affect policy-makers when deciding on different issues may lead to idealization and impractical recommendations. In the WWViews method citizens, unlike policy-makers, are considering exclusively the topic of climate change without necessarily having to deal with the consequences of their recommendations. This might result in a tendency towards rather radical recommendations since those making them do not have to consider all the ways in which they may affect society as a whole.

The novelty of the WWViews initiative together with its innovative character made the project a big challenge and an opportunity. It fitted perfectly with Simurg's profile and interest, but it was also a challenge as the group had only just been formally established. However, its close and strong links with the University made a huge difference, as mentioned above. The risks of being such a new organization were turned into a potential and a core strength, as its members put all their energy into making this project a big success.

The political environment around the issue of climate change was not ideal. Local funds were not available; the issue was on the political agenda but different aspects of it were scattered around several institutions and no clear counterpart was identified as its partner in the project. The Danish Embassy in Brasília funded the project. Yet, at the national level, support came from local government and a representative in Parliament. These two actors were key in complementing and supporting Simurg's efforts. It is important to note that the individuals behind the institutions that clearly committed to support the project had a special interest in both the policy issue and the deliberative process.

This project addressed a fundamental issue in the country that is likely to become more urgent in the years ahead. It contributed to broadening the scope of environmental discussion while also embedding it into society. Participants, facilitators, project organizers and a wide public became actors in the arena of environmental policy for the first time in our history.

Despite the limited impact of WWViews at COP15, and the failure of the climate convention to achieve its goal of a successor agreement to the Kyoto Protocol, we are heartened by the advance in participatory technology assessment occasioned by WWViews Uruguay. Our basic goal is therefore affirmed: we need to democratically engage the public in the most systematic way possible, not only to deliberate on S&T issues but also to be enabled to influence decision-making processes.

Acknowledgements

To the Danish Board of Technology and the Danish Embassy in Brasília. To the University of the Republic, the Science and Development Unit at the School of Science, to Amílcar Davyt and to our colleagues from Simurg.

Notes

1   The authors of this work were involved in the organization of WWViews Uruguay either as professors at the University or through Simurg.
2   The crisis over the traditional view of S&T became evident over the last decades of the 20th century. The academic reaction that weakened the hegemony of logical empiricism in the philosophy of science and the critical social reaction to technocracy converged in the proposal of the science, technology and society (STS) studies (Fuller, 2006).
3   The Rector is elected by the students', professors' and graduates' organizations, which contributes to the accessibility of the office.
4   After determining that the project was of interest, the Danish Embassy in Buenos Aires forwarded the proposal to its counterpart in Brasília, which handles environmental concerns in South America and ultimately funded the project.
5   The exit survey differed from one administered in 15 other WWViews countries. It called for 'agree/disagree' responses to a series of statements (respondents could also tick 'don't know / don't respond' options and add comments), aiming to mitigate the bias towards positive answers in surveys that offer a spectrum of positive and negative responses.
6   It would be interesting to evaluate other science communication initiatives being carried out in Uruguay and compare the results. Although Uruguay is gradually evolving towards more modern models of science communication, most of these initiatives are still based on outdated one-way models aimed at improving scientific literacy in what is perceived as a cognitive deficit among the population at large. A recent opinion survey (ANII, 2008) carried out on Public Understanding of Science, Technology and Innovation, for example, showed that despite the very positive ideas with which the term 'science' is mostly associated (knowledge and progress), most people perceive it as an activity that is remote from everyday life and is carried out by people who are innately disposed towards this activity, as if they were born with a scientific mind.

References

ANII (Agencia Nacional de Investigación e Innovación) (2008) 'Encuesta de percepción pública sobre ciencia, tecnología e innovación, principales resultados', ANII, Montevideo

Cerezo, J. A. L. and Cámara, M. (2007) 'Scientific culture and social appropriation of the science', *Social Epistemology*, vol 21, no 1, pp69–81

Davyt, A. and Lázaro, M. (2009) 'Da Teoria à Práxis: A evolução dos cursos sociais e humanísticos numa faculdade de ciências exatas e naturais', *Proceedings of the III SIMPÓSIO*, Nacional de Tecnologia e Sociedades: Desafios para a Transformação Social, Curitiba, 10–13 November

Decree of the Executive (20 May 2009), 1st article

Einsiedel, E. F. and Eastlick, D. L. (2000) 'Consensus conferences as deliberative democracy', *Science Communication*, vol 21, no 4, pp323–343

Fiorino, D. J. (1990) 'Citizen participation and environmental risk: A survey of institutional mechanisms', *Science, Technology & Human Values*, vol 15, no 2, pp226–243

Fuller, S. (2006) *The Philosophy of Science and Technology Studies*, Routledge, London

Funtowicz, S. O. and Ravetz, J. R. (2000) *La Ciencia Posnormal: Ciencia con la Gente*, Icaria, Barcelona

González García, M., Todt, O., Gutiérrez, I., López Cerezo, J. A., Estévez, B. and Luján, J. L. (2006) 'Participación pública en ciencia y tecnología', in J. Sebastián and E. Muñoz (eds) *En Radiografía de la investigación pública en España*, Biblioteca Nueva, Madrid

Haro, S. (2003) 'Periodismo Científico. Primera Aproximación a las Razones de su Escaso Desarrollo en Uruguay', PhD thesis, University of Montevideo, Uruguay

Hirschfeld, D. (2010) 'Uruguay hará juicio ciudadano sobre energía nuclear', *SciDev.Net*, 10 June 2010, www.scidev.net/en/climate-change-and-energy/nuclear-energy/news/uruguay-to-hold-public-trial-on-nuclear-energy.html, accessed 7 February 2011

Jasanoff, S. and Martello, M. L. (2004) 'Knowledge and governance', in S. Jasanoff and M. Martello (eds) *Earthly Politics: Local and Global in Environmental Governance*, MIT Press, Cambridge, MA

Lázaro-Olaizola, M. L. (2009) 'Cultura científica y participación ciudadana en política socio-ambiental', PhD thesis, Universidad del País Vasco, España

Martin, A. and Sherington, J. (1997) 'Participatory research methods: Implementation, effectiveness and institutional context', *Agricultural Systems*, vol 55, no 2, pp195–216

Merton, R. K. (1973) *The Sociology of Science: Theoretical and Empirical Investigations*, University of Chicago Press, Chicago

Wachelder, J. (2003) 'Democratizing science: Various routes and visions of Dutch science shops', *Science, Technology & Human Values*, vol 28, no 2, pp244–273

11

Adapting Citizen Consultation in a Small Island Context: The Case of Saint Lucia

*Karetta Crooks Charles, Jacquelyn Pomeroy
and Richard Worthington*

As one of only a few island nations and the only representative from the Caribbean, Saint Lucia played a unique and vital role in the WWViews project. Small island countries are the most likely to experience severe climate change damage. As a leader in the Caribbean, Saint Lucia's government has taken a strong stance in the international arena while making climate change response one of its top development priorities. However, the implementation of policies to avoid or adapt to climate change impacts has been slowed by various institutional, financial and practical obstacles. Consequently, WWViews in Saint Lucia, and presumably other small island states, may prove more effective as a tool to expand public awareness and generate grass-roots approaches to adaptation efforts, rather than as a way for the people to call for urgent action on the part of the government. In this chapter, we describe Saint Lucia's unique natural, economic and political setting, review the process and outcomes of the WWViews process, and explore the potential of citizen participation to support sustainability policies and practices in a small island context.

Climate change: Nature and society in Saint Lucia

Saint Lucia is a country of 167,000 people situated on a small, mountainous, volcanic island in the eastern Caribbean. Its initial Ciboney and later Arawak inhabitants were replaced over time by Caribs prior to European contact, after which Dutch, French and British invaders contended with the Caribs and one another for more than three centuries before the British gained lasting control of the island early in the 19th century. A century of British rule ensued, yielding to progressively expanding self-governance over six decades that culminated in full independence from the UK on 22 February 1979 (Tulsie et al, 2001, pp18–19).

Annual income was $9020 per capita in 2010, placing Saint Lucia in the World Bank's 'upper middle income' category.[1] Children spend an average of 14 years in school, and more than 90 per cent of the population is literate (CIA, 2010). One indicator of the government's legitimacy is Transparency International's *Corruption Perceptions Index* (2009), on which Saint Lucians' perceptions of their public sector ranked 22nd in the world, behind only Canada, the US and Barbados in the western hemisphere. However, even with stable, transparent governance and moderate income and education levels, Saint Lucia faces significant economic challenges, as reflected in rates of unemployment and poverty at 20 per cent and 25 per cent, respectively. With 47 per cent of GDP generated by tourism, the economy remains externally oriented, as reflected in a 2008 government report that announced plans for an additional 7000 hotel rooms over the following decade (Ephraim, 2009, p10). The agricultural sector (a legacy of the plantation era) is in decline, but nonetheless employs nearly 22 per cent of the workforce (CIA, 2010).

The largest threat to the country's future is posed by the interaction between these socio-economic systems and a changing global climate. Numerous scientific studies have established that the Caribbean – and small islands in general – face a dire climate threat compared to the rest of the world (Challenger et al, 2001; Mimura et al, 2007). Not only are islands more vulnerable to the effects of climate change such as sea-level rise, higher land and water temperatures and violent meteorological events, but they are also generally less well equipped to mitigate and adapt to the coming risks because of their distinctive geographical and socio-economic characteristics. These challenges directly influence the development capacity and quality of life in small island developing states (SIDS) because adaptation efforts are expected to take up a disproportionately high share of GDP compared to developed nations (Bueno et al, 2008).

The main impacts of climate change in Saint Lucia are expected to be higher sea levels, seasonal fluctuations in precipitation and temperature, and extreme weather events (Table 11.1). With a land mass substantially comprised of rugged mountain landscapes, and a post-contact history of enclave industries created by external powers, both geography and history have combined to concentrate settlements and infrastructure in the flat and accessible coastal regions. Coastal wetlands, ports and docks, low-lying farmland, beaches and the resorts that benefit from them – in essence, Saint Lucia's economy and society – therefore face serious threats.

Public understanding of climate change by Saint Lucians is characterized by insightful observations of local phenomena and a limited understanding of their larger context. In a 2006 research project by Saint Lucia's environment ministry (Sustainable Development and Environment Section, 2006), nearly 90 per cent of respondents to a survey, many of them with little education, reported observing changes in weather patterns over the previous 10 years.[2] Of these respondents 20 per cent noted a change in the timings of wet and dry seasons, essentially confirming what scientific studies on changing precipitation patterns in the Caribbean have shown (Mimura et al, 2007, p691). Nonetheless, the survey revealed that the larger context and implications of

Table 11.1 *Dimensions of climate change impacts in small island states*

Dimensions	Climate change impacts
Size and settlement dynamics	Small land base and (in many cases) high population growth and density promote: • intense exploitation of local resources for subsistence and export; • import of necessities not locally available (e.g. fossil fuels); • concentration of population near high-risk areas prone to storms or flooding.
Economic isolation and dependence	Origins as colonial outposts, small land base and limited economic diversity create: • reliance on one or two sectors (tourism and plantations in Saint Lucia) more susceptible to abnormal climate variation than industries that are less dependent on natural resources; • long distances to primary markets; • sensitivity to external market shocks.
Ecological vulnerability	Small islands' exposure to predicted increase of catastrophic weather events is a severe threat because: • many (including Saint Lucia) are in the path of hurricanes; • thin freshwater lenses are subject to salinization from sea-level rise and storm incursions, harming potable water and soils; • telecommunications, ports, energy and tourism are close to the coastline; climate events can turn a net exporter into an importer for up to two years, exacerbating economic vulnerability.
Infrastructure	Temperate climates and weak economies have produced housing and other infrastructure poorly equipped to withstand more frequent and intense weather events.

Source: adapted from Mimura et al (2007); Sustainable Development and Environment Section (2009); CIA (2010)

climate change were poorly understood. For example, most respondents had heard of climate change and two-thirds believed that Saint Lucia would be affected, but nearly 60 per cent did not think it would affect their work (p9).

Notwithstanding the active and outspoken role of national leaders in the United Nations Framework Convention on Climate Change (UNFCCC) and other international venues, climate change receives little coverage from local media, and active engagement with the issue is limited to the handful of government and industry experts supported primarily by UN and other international sources. The citizens who will bear the impacts of climate change, therefore, have a limited sense of the threats looming on the immediate horizon.

Climate change policy: Global, regional and local

From a global perspective, the economic and political factors in Saint Lucia create a unique dynamic with interesting implications for WWViews as a method. Small islands produce only 1 per cent of greenhouse gas (GHG) emissions, but are disproportionately subject to the effects of climate change (Mimura et al, 2007, p690). The associated costs and benefits of strong international climate change policies therefore vary considerably from those of the developed nations that have a greater say in international negotiations. On one hand, the external economic dependence of SIDS places significant financial and political control over the mitigation and adaptation measures that might improve their future prospects in the hands of the very powers who account for the vast majority of emissions. The latter, on the other hand, face enormous costs to not only change their own ways, but to help finance global adaptation in peripheral regions and countries. They therefore face the prospect of paying the remediation costs for climate change impacts that will be most severe in other countries, while SIDS and least developed countries stand to receive investment and development opportunities in an international climate deal.

Accordingly, the Government of Saint Lucia, along with other Caribbean states, has taken a strong stance on GHG reduction targets, even in the absence of a vocal public demand for action. Climate change is the top priority for the Alliance of Small Island States (AOSIS), which lobbies for small islands within the UN. In the Caribbean, the Organization of Eastern Caribbean States (OECS), which is based in Saint Lucia, has taken on sustainable development as one of its main programme areas, and the chief economic organization in the region, the Caribbean Community (CARICOM), executed a Renewable Energy Development Program among its 15 member states and carried out an adaptation plan from 2004 to 2007.

Saint Lucia has been an important catalyst in this network and an outspoken voice in the UNFCCC. The country was among the earliest signatories to the UNFCCC in 1993, and in 1999 the national delegation to COP5 in Bonn declared their intention to become a Sustainable Energy Demonstration Country. At COP6 the following year in The Hague, Prime Minister Kenny Anthony formally announced a project to develop a national Sustainable Energy Plan (SEP) in partnership with the Climate Institute (Washington, DC), the Organization of American States, the United Nations Development Programme and the US Department of Energy, among others (Anthony, 2000, p5). At COP15, Prime Minister Stephenson King called for a 1.5°C cap in temperature rise (25 per cent below the 2°C cap advocated by most players that still awaits action), declaring that 'in life, and diplomacy, many things are negotiable, but the survival of our island homes is not' (King, 2009). This seemingly radical position is common sense from the point of view of Saint Lucia. King has no domestic constituencies – such as big energy corporations – that strongly oppose climate change efforts, but both near-term flows of capital to combat climate change, as well as the longer-range struggle to survive, provide compelling reasons to support an aggressive climate deal.

Saint Lucia's auspicious mix of clear incentives, an active leadership role, and stable and legitimate government nonetheless have yet to deliver significant results. According to a review of SEP, for example, 'a number of targets are outlined in the plan for delivery of electricity via renewable energy sources but eight years later none of those targets have been achieved' (Ephraim, 2009, p5). The factors that account for this include the large investment required for such a plan and the uneven flow of funds to date; regulatory frameworks suited to a monopoly electricity provider rather than the many players needed to implement conservation and renewable energy projects; isolation which, for example, makes it difficult to acquire energy-efficient appliances that are readily available at reasonable prices in major markets; and limited public awareness of climate change and energy alternatives.

WWViews Saint Lucia

On 26 September 2009, 67 citizens gathered in Castries, the capital of Saint Lucia, to share their views on global warming in the hope that their message would be passed on to national politicians and COP15 negotiators. As the only Caribbean nation to participate in WWViews, the Saint Lucia case offers important data for understanding the value of WWViews among SIDS.[3]

Saint Lucia National Trust and its WWViews partners

Like most non-governmental organizations (NGOs) with a mandate to conserve the environment, the Saint Lucia National Trust (SLNT) has a growing concern about climate change and its impending impact on SIDS and the rest of the world. Considering the failure of the Kyoto Protocol to meet its target and the gloomy projections by the Intergovernmental Panel on Climate Change, the Trust felt that the negotiations at COP15 should result in a strong binding agreement, which should be influenced by the people who will have to live with and learn to adapt to this phenomenon. Therefore, the Trust welcomed the Danish Board of Technology's (DBT) invitation to be the local organizer of WWViews.[4]

SLNT is a statutory body established in 1975 for the specific purpose of preserving buildings, objects of historic and architectural interest and areas of natural beauty together with their animal and plant life, for present and future generations. The Trust owns or has been vested with responsibility for parks, nature reserves and historical and archaeological sites, and provides recreational and educational opportunities within these areas for residents and visitors. SLNT is a non-profit organization and all revenues generated by grants, donations and user fees to its sites are used for the conservation of the natural and cultural heritage of Saint Lucia.[5] The Trust is headquartered at the Pigeon Island National Landmark, which attracts many visitors to its historical ruins, beaches, beautiful lawns and picnic facilities. Given its responsibilities and the fact that citizen participation has been mentioned in numerous studies as an important part of the climate change adaptation efforts essential to the preservation of Saint Lucia's natural heritage, the Trust viewed DBT's invitation as an important opportunity.

The SLNT has for many years partnered with the Sustainable Development and Environment Section (SDES) in the Ministry of Physical Development and the Environment (2003), which is also the national focal point on climate change. The Trust approached the SDES for financial and technical assistance to execute the project, and was successful in securing the latter. SDES also assisted in recruiting participants by sending invitations to all the groups in their network. In addition to the SDES, the Trust secured technical support from OECS to help with facilitation. One motivation in contacting OECS was to indirectly involve the wider OECS region and build awareness of Saint Lucia's participation in the project. Finally, the SLNT met with its Youth Group in order to brief them on the project and to request their assistance in its execution as staff or participants at the actual meeting.

A vigorous effort to raise funds for the WWViews consultation from about 20 companies yielded a contribution of EC\$3000 (about US\$1120) from the Bank of Saint Lucia. In addition to this contribution, several businesses donated prizes to entice participants to stay for the entire event. These included dinner and brunches at prestigious hotels, an inter-island cruise and assorted gifts from a prominent jewellery franchise.

Recruitment and staff

The recruitment process was one of the most significant challenges faced by WWViews Saint Lucia. The country is divided into four quadrants: Castries, Gros Islet, Vieux Fort and Soufriere. Considerable effort was expended to attract participants from all the regions and their surrounding districts.

After DBT approved the Trust's recruitment strategy, letters were sent to scores of organizations ranging from banks, supermarkets and teachers' associations to transport associations, farmers, fishermen groups and the churches, inviting them to spread the word among their staff, families, friends and communities. Press releases promoting the event were also sent to both print and electronic media. Brief notices were aired on *Community Diary* programmes on the various radio stations and an advertisement was placed on Radio Saint Lucia as well as an interview on that station's popular morning programme *The Agenda*. The internet was also utilized to get the message across as emails were circulated on several occasions to the over 1000 members of Saint Lucia National Trust.

Another important strategy employed was on-site recruitment in three of the targeted geographic locations of the island; Vieux Fort, Soufriere and Gros Islet. There were two members of staff at the first two locations. While one stayed on location the other traversed the streets handing out invitation letters and encouraging Saint Lucians to sign up. A booth was mounted at a popular intersection frequented by the local population, and passers-by were invited to sign up. Invitation letters were also handed out to pedestrians and various entities including hairdressing salons, travel agencies and the local market in the capital city. It was anticipated that this method would significantly help the recruitment process; however, instead of displaying an interest in the project, many passers-by were more eager to find out if something were being distributed free. Overall, the recruitment effort achieved more than the 100 necessary registrations, but on WWViews day only 67 participants showed up, a 33 per cent rate of attrition that is significantly higher than the 20 per cent rate expected by DBT for deliberative events that it conducts in Denmark. Furthermore, the turnout from the more rural areas of Soufriere and Vieux Fort and their environs was limited.

While different recruitment strategies might have yielded a more representative sample of Saint Lucians, the outcomes may also reflect that Saint Lucia joined the WWViews Alliance late in the project. On a positive note, the participants ranged in age from 18 to over 70 and several levels of university graduates were represented as well as those who had only completed secondary school. Some of the occupations present were tour guides, bank tellers, teachers and engineers. There were also the unemployed and some retirees. Although detailed demographic data was not collected, it is likely that the inability to pay expenses and support special outreach efforts (e.g. conducting targeted recruiting for the segment of the population that speaks only Kweyol) resulted in participants who were more educated and presumably wealthier than the average Saint Lucian. This also could have been influenced by the use of SLNT networks to recruit participants. One would assume that this population is more concerned and aware about environmental issues on average. One indication of this is that 33 per cent of participants said that they 'knew a lot' about climate change and its consequences before joining WWViews, compared to 28 per cent of global WWViews participants and the 6.5 per cent of the 2006 survey in Saint Lucia who said they were 'very knowledgeable' about the environment (see note 2 regarding the low education levels of respondents to the last survey).

The WWViews staff totalled 19 people representing categories of institutions that we list in parentheses: ten were employees of the SLNT (NGO staff), three were Trust members (engaged citizens), five were from SDES (government agency), and one was from the Environmental and Sustainable Development Unit within the OECS Secretariat (regional organization). Letters were sent to prospective candidates and organizations which the Trust felt would do an efficient job in executing their duties as WWViews staff. The letters as drafted in the WWViews Manual outlined in detail the expectations and roles of the various staff, from the programme manager and group facilitators to the intranet reporters and vote counters. As a result of the staff's hard work the meeting was a success; however, it must be noted that three of the group facilitators who had given their commitment to the project fell out almost at the last minute, thus reducing the number of facilitators to eight. As a consequence, two citizens who possessed the necessary skills to serve as facilitators switched to the latter role in order to fill the gap. Noting the diverse types and levels of organizations represented among project staff, the collaborative approach to conducting Saint Lucia's first deliberative exercise can itself be viewed as a milestone of inclusive governance.

Results and outcomes

The event itself went according to plan, with the results being reported in real time through the WWViews web tool for immediate processing and posting on the internet. Among the many comments indicating the good quality of deliberation were those by a facilitator who observed that 'the discussion in my group was very interesting, and reflected a variety of perceptions, varying knowledge and interests in the topic of climate change. It was interesting to see members teaching each other, and the group deliberating to find common ground'.[6]

The participants called for more urgent action to address climate change than their cohorts worldwide, suggesting an awareness of the heightened vulnerability of SIDS. The Saint Lucia responses to items about the significance of climate change and the urgency of responding to it clearly expressed greater concern than global responses to the same questions. An overwhelming 89 per cent of Saint Lucian participants said they were 'very concerned' about the issue compared to only 62 per cent of the world participants; 99 per cent versus 91 per cent said that a climate accord 'is urgent, and a deal should be made at COP15'; and 96 per cent urged their national leaders to give 'high priority' to any climate deal made at COP15, compared to 91 per cent worldwide.

At all the WWViews sites, questions about controlling emissions and enhancing adaptation elicited responses that were more evenly distributed across the range of choices than those for questions addressing the urgency of the issue and the need for action. This is consistent with research on other deliberations, which shows a similarly even distribution and granularity in views on the particulars of practical policy actions (Fung et al, 2008, p11). Within this pattern of a more even range of responses, however, Saint Lucians nonetheless stood out in comparison to other country categories. For example, a full 50 per cent of Saint Lucians said that a short-term greenhouse gas emissions reduction target for Annex 1 countries[7] should be 'higher than 40 per cent', the most stringent option, whereas just 31 per cent of global respondents and 35 per cent of respondents from the middle-income group of countries that includes Saint Lucia supported this option. Saint Lucians were just below the global percentage of respondents in expecting rapidly growing and/or high-emissions countries to meet the same targets as Annex 1 countries, but favoured requiring lower-income countries to meet these standards by a slightly greater margin than most other countries. Overall, Saint Lucians favoured the most stringent choice on these three items by margins that exceed the responses of other categories of countries by 24 to 41 per cent (see Table 11.2). In addition to the even greater concern about climate change registered by WWViews Saint Lucia participants compared to their counterparts around the world, the results show a higher propensity to advocate strong action at international level.[8]

Table 11.2 *Percentage of respondents favouring strictest emissions reduction targets*

Response (most stringent choice for each item, 5 choices per question)	Saint Lucia	Middle-income countries	Low-income countries	Annex I countries	World
'Short-term reduction targets for Annex I countries should be higher than 40 per cent'	50	35	29	28	31
'Non-Annex I countries with substantial income and/or high emissions should meet the same targets as Annex I countries'	25	25	26	27	27
'Low income countries should meet the same targets as Annex I countries'	18	15	17	11	13
Amount that Saint Lucia response for above three items exceeds other groups of countries	N/A	24	29	41	31

Ultimately, participants, facilitators and policy-makers offered both praise and criticism of the WWViews project. The head of SDES, Crispin d'Auvergne, has long held that it is absolutely critical for citizens' views to be incorporated into the policy-making process and that WWViews gave citizens an opportunity to do so. A younger participant in her 20s from Soufriere said, 'we live in a democracy in Saint Lucia and I'm of the opinion that the citizens' voices are not being heard enough and our politicians need to get more actively involved in the community and hear what they [the citizens] want'. A facilitator agreed that the decision-makers can benefit, yet expressed the criticism that it must be done on a larger scale and be given some political significance.

The facilitators expressed disappointment that the modest Copenhagen Accord (UNFCCC, 2009), which was the main outcome of COP15, did not reflect the recommendations of WWViews. On the other hand, the Saint Lucia delegation to COP15 did support policies in line with their citizens' viewpoints. Prime Minister King received the WWViews results prior to speaking at COP15, and some of the views he expressed reflected the results and conclusions of the WWViews participants, giving legitimacy and credence to the exercise and its outcomes. One participant expressed this as follows: 'We should definitely have similar meetings on a whole range of topics. The political decision-makers can certainly benefit because when they go to speak at international fora they know they have the opinions and voice of the people behind them so they can speak with more conviction.'

A facilitator elaborated on this theme in her observation that leaders can have confidence that the views represented informed debate among the participants rather than superficial reactions of the sort often reported in polls or media coverage. In the facilitator's words:

> *Decision-makers need to be sensitized on these pressing issues. The consultation allowed for discourse among persons of different interests, backgrounds, experience, among other things, and provided opportunities for persons to learn from each other, to flesh out issues, and to reach common ground in an environment tailored to achieve the project's objective.*

These comments from participants and facilitators clearly suggest their desire to move beyond the current posture in which Saint Lucians generally recognize the legitimacy of the Government of Saint Lucia to a more active and sustained engagement between citizens and leaders.

Deliberation, participation and development in a small island context

In this concluding section we will discuss the following questions: What did WWViews Saint Lucia accomplish in the domain of deliberative democracy? What are the potential contributions that deliberative and participatory practices can make towards effective climate change and development policies in Saint Lucia, as well as SIDS more generally? Finally, what constraints and resources are likely to shape future efforts to realize these potentials?

Accomplishments

WWViews Saint Lucia can claim three important accomplishments. First, the deliberation and dissemination of results to policy-makers clearly met the primary WWViews goal of providing ordinary citizens the opportunity to influence global climate policy. The first-ever global policy deliberation was also the first citizen policy deliberation in Saint Lucia, making this an even more important milestone. Second, valuable insight for future deliberative events was gleaned from participants and facilitators during and after WWViews day. In addition to affirming the importance of informed and considered public deliberation to at least some Saint Lucians, this input advocated direct and sustained engagement between citizens and politicians, and extension of participatory practices to other policy issues.[9]

The most important accomplishment of WWViews Saint Lucia was the organizational development that it produced. SLNT built on a strong relationship with the environment ministry to draw together a diverse and multilevel project team and ancillary relationships for the project. At the regional and international levels, this included the direct participation of OECS in the Saint Lucia deliberation, as well as SLNT's engagement with the WWViews Alliance. The horizontal connections with project managers from Cameroon and the Maldives that were made through the training seminar are especially noteworthy, as these types of solidarities are critical to any prospect for global environmental governance. However, it is difficult to activate these ties in interstate relations and especially in civil society relations, because the enduring legacy of colonialism dominates international interactions for post-colonial nations (e.g. through commercial and educational ties to the colonizing country), and the resources needed to break out of this pattern are constrained, especially for NGOs.

The external WWViews network was complemented by inward networking with SLNT's Youth Group, members and the private companies and civic groups (e.g. teachers' union) who were asked to support the project. The Government of Saint Lucia has recognized the importance of such collaborative relationships as well as increased citizen participation in its sustainability and climate change efforts since the 1990s, but progress has been modest at best. One spin-off of WWViews, however, was the adoption of participatory methods in a government energy symposium in the months after the climate change deliberation. In bringing together diverse organizations for WWViews in a civil society context, SLNT advanced the citizen participation agenda considerably.

Potential

The most immediate potential of WWViews is improved future deliberations that engage a broader cross-section of the population and generate sustained engagement of citizens and policy-makers. The environment ministry's 2006 survey showed that Saint Lucians were unsure how climate change might affect their nation and livelihoods. After deliberating the issues, however, the WWViews Saint Lucia participants registered a sense of urgency and need for action that exceeded even the high levels voiced in other nations. This underscores the potential for WWViews as a tool to raise public awareness and encourage engagement with politicians. Changes that can strengthen the existing model include:

1   customizing educational materials to the specific circumstances of Saint Lucia, the Caribbean and SIDS;
2   better focusing recruitment by developing targeted outreach strategies for marginalized populations such as Kweyol speakers and working more closely with leaders among marginalized groups by involving them in planning the event;
3   developing strategies to sustain engagement with and by the participants.

In general there is a need to develop local capacity to design, conduct, assess and report on these events if the methodology is to become a mainstream approach to shaping public policy. Building partnerships with community groups can produce a more positive response and higher levels of participation, but this will take time because deliberation is new to Saint Lucia. Should the opportunity arise, continued participation in the

WWViews Alliance can be an important means of building on the accomplishments of SLNT. All these improvements would require additional resources, although to some extent reallocating effort might reduce the added burden.

Beyond additional WWViews events, the calls for more active engagement between policy-makers and citizens shift attention to the origins, purposes and wider potential of both deliberative and participatory practices. An important influence in the WWViews method is the participatory approach to technology assessment pioneered by DBT and others, which arose in order to overcome the exclusion of citizens from seemingly obscure technological decisions that yield major effects as they permeate society over time. But climate change represents the outcome of technology decisions made over the past century rather than a threat from ideas that are in the early stages of development. This renders the notion of early intervention moot, and the incentives for SIDS leaders to be vocal advocates in the UNFCCC makes a citizen voice less urgent than it is in countries where these issues are hotly contested.

Rather than informing leaders about citizen perspectives on global policy, the most significant opportunities for citizen engagement in SIDS could instead be policy discussions and practical mobilization for the adaptation that lies ahead. One study has projected that, in the absence of effective adaptation strategies, the cost of climate-induced damage could be as high as a quarter of Saint Lucia's GDP by 2050 (Bueno et al 2008, p14). Even a much lower impact could pose changes similar to the shift from agriculture to tourism that has occurred in recent decades. If international agreements result in increased flows of funds to SIDS and other vulnerable countries in the decades ahead, participatory adaptation could become a development model as well as a defence mechanism, engaging citizens in an active mode rather than trying to direct them from above. The participatory toolkit would not only include deliberations on adaptation policies in Saint Lucia and other SIDS, but also popular education that helps ordinary people connect impending environmental and economic changes to their own lives, community-based research to observe changes and experiment with responses to them, and participatory planning for community protection and development. A concrete case in which participatory adaptation can be more effective than top-down strategies is that of fishers, who face the challenge of learning how fish have adapted to climate change (Challenger et al, 2001). If fishers are actively engaged in generating this knowledge, they will then be in a good position to determine what type of fishing industry (if any) the remaining stocks can support, how the existing industry needs to adapt in order to harvest the fish on a sustainable basis, and what strategies can be followed to accommodate those who are displaced by these changes.

Constraints and resources

The limitations on increased citizen participation in Saint Lucia are clear but significant: as in many countries, formal democracy has developed more rapidly than participatory practices, and a high poverty rate consigns much of the population to a marginal and poorly informed place in public discourse, as demonstrated by the KAP survey (Sustainable Development and Environment Section, 2006). Perhaps because of the relative disinterest among the public, the media have paid little attention to climate change.

Alternatively, numerous factors could turn out to be important in expanded participation:

- The threat of climate change is clear and immediate; expanding awareness should be less challenging than in countries where there is less clarity.
- There are no significant domestic political forces opposed to addressing the issue.
- Education levels are fairly high and will increase following reforms in the past decade.
- People who receive information and have the opportunity to discuss it (WWViews participants) see climate change as urgent, advocate strong international measures and call for increased citizen engagement.
- The Government of Saint Lucia is a consistent supporter of sustainable energy development and strong climate change policy.
- An organization with an appropriate mission and stable funding base (SLNT) can play a leadership role.
- Participatory processes are comparatively inexpensive, but arguably critical in mobilizing the country's human resources for the challenges that lie ahead.
- There is a reasonable prospect that capital flows for adaptation will increase.

It remains to be seen whether these potentials can be turned into effective action. If they do bear fruit, however, the benefits can reach far beyond the small population that is directly involved: as Saint Lucia and other SIDS are literally going under water, they might be able to point the way to higher political ground.

Notes

1 This is a 'purchasing power parity' estimate for 2008 that adjusts the income expressed in US dollars to its purchasing power in the country being measured ('World Bank – Country Classifications' at http://data.worldbank.org/about/country-classifications, accessed 7 August 2010).
2 The survey was conducted in the Vieux Fort region of the country where resource-based activities such as farming and fishing predominate. Over half (52.1 per cent) of respondents had primary education or less (p7), whereas most children in Saint Lucia now finish secondary school (CIA at www.cia.gov/library/publications/the-world-factbook/geos/st.html, accessed 6 January 2011).
3 The Maldives was also a WWViews partner, and participated in the same project manager training seminar as Saint Lucia (along with Cameroon) that was convened for those who joined the project too late for the main seminar held in March 2009. This provided a rare

opportunity for horizontal communication below ministerial level by concerned professionals from developing countries.

4   The Trust's commitment to a particular position on climate policy and its close association with a vocal national government raise legitimate questions about its neutrality as a WWViews partner. However, several factors might be cited in response to such concerns. First, the government's advocacy for strong climate change policies has been consistent over time, but it has been voiced by competing political parties that occupied the majority position in Parliament during different periods since 1993. This position can therefore be considered to be reflective of the range of views in the country. Second, in a small country with no university and few NGOs, there is no practical alternative to the Trust as the project manager. The choice is therefore to have a deliberation conducted by an 'interested party', or to exclude Saint Lucia altogether. Finally, the Trust was required to comply with all the requirements for WWViews partners and was subject to oversight by DBT, both of which establish a check on (if not a guarantee against) inappropriate advocacy.

5   A quarter of the Trust's approximately US$700,000 annual budget is provided by government subsidy and the remainder is raised from user fees, membership fees, grants and donations.

6   Two participants were interviewed by K. Charles at the end of WWViews day, and three facilitators were interviewed after the COP15 meeting in December 2009.

7   Annex 1 countries are the 39 Parties listed in the first annex to the UNFCCC, which lists the most affluent countries in the world.

8   WWViews results are posted at www.wwviews.org/node/287 (accessed 5 January 2011).

9   The wider significance of these important accomplishments is nonetheless modest for at least two reasons. First, the small size and unique position of SIDS in global climate politics make access to politicians more direct and less complicated than in most countries, even though few channels actually exist for such communication. Russia provides a stark contrast: it is large, has little experience of democracy, and is a major fossil fuel exporter, so any signs there of policy-maker interest in WWViews would be far more significant. Second, the value of the citizen feedback on the Saint Lucia deliberation is mitigated by the limited number of research interviews that could be completed within the scope of this project.

References

Anthony, K. (2000) 'Small island states most vulnerable to climate change impact: Feature address by the Prime Minister of Saint Lucia, the Right Honourable Dr Kenny Anthony', 6th Conference of the Parties to the UNFCCC, The Hague, The Netherlands, 13–24 November 2000, www.iisd.ca/climate/cop6

Bueno, R., Helzford, C., Stanton, E. A. and Ackerman, F. (2008) *The Caribbean and Climate Change: The Costs of Inaction*, Global Development and Environment Institute, Tufts University, USA, http://ase.tufts.edu/gdae/Pubs/rp/Caribbean-full-Eng-lowres.pdf

CIA (2010) *The World Factbook*, Central Intelligence Agency, Washington, DC

Challenger, B., John, L. and Raynold, J. (2001) *Saint Lucia Climate Change Vulnerability and Adaptation Assessment*, Government of Saint Lucia, Castries

Ephraim, J. (2009) 'Sustainable energy development in Saint Lucia: A discussion paper', Sustainable Development and Environment Program, Ministry of Physical Development,

Environment and Housing, Government of Saint Lucia, Castries, www.bb.undp.org/uploads/
file/pdfs/energy_environment/Discussion%20Paper%20Sustainable%20Energy%20in%20
St%20%20Lucia%20by%20J%20%20Ephraim.pdf

Fung, A., Lee, T. and Harbage, P. (2008) *Public Impacts: Evaluating the Outcomes of
CaliforniaSpeaks Statewide Conversation on Health Care Reform,* America*Speaks*,
Washington, DC

King, S. (2009) 'Address to the High Level Segment of the 15th Conference to the United
Nations Framework Convention on Climate Change by the Prime Minister of Saint Lucia',
16 December, www.pm.gov.lc/statements/2009/address_to_the_high_level_segment_of_
the_15th_conference_to_the_united_nations_framework_convention_on_climate_change_
by_the_hon_stephenson_king.htm, accessed 7 August 2010

Mimura, N., Nurse, L., McLean, R. F., Agard, J., Briguglio, L., Lefale, P., Payet, R., and Sem,
G. (2007) 'Small islands', in M. L. Parry et al (eds) (2007) *Climate Change 2007: Impacts,
Adaptation and Vulnerability: Contribution of Working Group II to the Fourth Assessment
Report of the Intergovernmental Panel on Climate Change*, Cambridge University Press,
Cambridge

Ministry of Physical Development, Environment and Housing (2003) *Saint Lucia National
Climate Change Policy and Adaptation Plan*, Government of Saint Lucia, Castries

Sustainable Development and Environment Section (2009) *Saint Lucia Climate Change
Vulnerability and Adaptation Assessment – Coastal Sector*, Government of Saint Lucia,
Castries

Sustainable Development and Environment Section in the Ministry of Physical Development,
Environment and Housing (2006) *Report of the Knowledge, Attitude Practice (KAP) Survey
for the Vieux Fort Region in Fulfillment of the VCA Project*, Government of Saint Lucia,
Castries

Transparency International (2009) *Corruption Perceptions Index 2009*, www.transparency.org/
policy_research/surveys_indices/cpi/2009/cpi_2009_table, accessed 31 July 2010

Tulsie, B., d'Auvergne, C. and Barrow, D. (eds) (2001) 'Saint Lucia initial national
communication on climate change', Ministry of Planning, Development, Environment and
Housing, Government of Saint Lucia, Castries, http://unfccc.int/resource/docs/natc/lucnc1.
pdf

UNFCCC (United Nations Framework Convention on Climate Change) (2009) 'Copenhagen
Accord', Copenhagen, http://unfccc.int/resource/docs/2009/cop15/eng/l07.pdf, accessed 18
December 2010

Part V
Studying Policy and Media Impacts

12

WWViews and Lobbying in Finnish Climate Politics

Mikko Rask and Maarit Laihonen

WWViews is a new concept for deliberative democracy at both national and international levels of climate policy. Despite its status as the first-ever global citizen consultation, WWViews has simultaneously been introduced and adopted in many countries and policy cultures beyond its original context, Denmark. Even though some aspects of cultural variation were taken into account in the design of the WWViews concept, especially through international partners who contributed to its planning, it is still relevant to ask whether the assumptions of WWViews concerning citizen participation in policy-making transfer well.

Finnish COP15 delegates – as well as delegates from all other negotiating countries – received demands, proposals and comments from various societal actors before COP15. What was and is the significance of these comments from outside official politics to people who work with national and international climate policy? What about in the future – could stronger empowerment of civil society lead to more acceptable policy? Can deliberative input be understood in the context of traditional lobbying, or does it require a new model of interaction between citizens and officials? This chapter studies traditional and novel lobbying[1] through attempts to use WWViews as a tool for impacting debate and policy on climate change in Finland. We note that the unfamiliar status of WWViews makes it a challenging case to 'sell' to policy-makers and the public. We analyse the factors that enhance or inhibit the local policy impact and transfer of WWViews. The question is topical, because the WWViews concept has raised increasing interest among WWViews Alliance countries and beyond. Furthermore, effective transfer of the WWViews concept to different national contexts will be necessary if future WWViews and other international citizen deliberations are to achieve their objectives of increased democratic participation by citizens and impact on political decision-making (for the objectives of deliberative processes, see e.g. Lidskog and Elander, 2007; Renn, 2008).

We study the transfer of the WWViews concept from two perspectives: the attractiveness of the concept to policy-makers and other national climate policy actors including media; and how the concept has been tuned to the Finnish context, i.e. which factors have been retained or modified. Adjusting foreign concepts to national contexts is necessary and should include a comprehensive consideration of relevant local issues; our particular interests are in Finnish lobbying strategies, which include dissemination and cooperation efforts before, during and after the WWViews event.

We analyse the lobbying efforts and policy transfer of WWViews through two theoretical perspectives: first, the literature on policy transfer, especially the advocacy coalition framework (ACF) originally developed by Sabatier (1988) that we use as a heuristic model for studying relevant factors of policy transfer; and, second, a model of stakeholder and citizen participation (Renn, 2008, pp290–304) that provides a framework for identifying different objectives, rationales and modes of participation considered by different policy actors as most appropriate and justified.

As empirical data we use policy-maker interviews that focus on the strengths and weaknesses of WWViews and similar exercises in citizen participation in the Finnish policy context; data on the media coverage of WWViews in Finland; and participant observations from various project meetings and negotiations.

The case of Finnish WWViews provides evidence for factors that enhance or inhibit the transfer of innovative approaches to deliberative democracy and citizen participation in climate policy. What conditions are required by the policy actors for a successful adaptation of WWViews and other similar deliberation processes? What design aspects should be taken into account to make future WWViews more easily adapted to particular national contexts?

Finnish background and lobbying

The Finnish system of policy-making has often been characterized as elitist, expert-driven, and embedded in a techno-optimist and consensual environment (Rask, 2009). In his analysis of Finnish technology policy culture, Pelkonen (2008, pp62–63), for example, has claimed that the decision-making system has remained rather closed with respect to civic organizations and especially the issue of public participation. He has characterized the Finnish system as 'exclusive corporatism' as distinct from more inclusive corporatist systems as in Norway (Kallerud, 2004) or from more deliberative and participatory systems as in Denmark (Bertilsson, 2004). Some recent studies also suggest that the majority of Finnish authorities do not trust the lobbying of civic organizations (Rantanen, 2009). In terms of participation, Savikko and Kauppila (2009, p4) claim that in Finland, as a result of trust in experts and the consensual culture, participation is driven by authorities and favours participation by established actors through working groups and committees. Against this cultural background, it is relevant to ask how attractive Finnish policy-makers consider the striving for more direct civic influence to be.

WWViews represents a less familiar paradigm in Finnish policy culture and therefore presents additional challenges for lobbying. Even though it utilized a media strategy that consisted of some well-known tools (e.g. press releases, direct contact with politicians, blogging), the project experienced difficulty in gaining policy attention. As the UN conference was approaching, a large number of other projects were also aimed at national policy-makers: engineers created their own climate programme and civic organizations banded together (*Järjestöjen kymmenen vaatimusta Kööpenhaminen ilmastokokoukselle*, 2009; TEK and UIL, 2009).

As WWViews in Finland was independent of traditional policy actors such as political, industrial and environmental organizations, it has an interesting standing among customary actors who also struggle for the attention of policy-makers. Traditional actors have established status in lobbying in politics and today climate issues are perhaps one of the most popular lobbying focuses. WWViews aims to bring out the voice of an ordinary citizen, but it faces the challenge of how to grasp policy-makers' attention when they are being effectively lobbied by a number of other interest groups with better direct access.

Lobbying is often filled with conflicts as different actors aim to raise their views and desires concerning future policies. Tensions also occur between parties that work on the 'same side'. The conflicts and tensions, however, should be seen not only as a constraint, but also as a source of creativity. Another important element of lobbying is cooperation: creating alliances between actors enhances the success of the lobbying process (Aitken-Turff and Jackson, 2006). These are challenges and possibilities especially for WWViews and similar novel models that aim to enter – and make an impact in – the arena of climate policy.

Ambiguous reception of WWViews in Finland

In Finland, WWViews was organized by the National Consumer Research Centre (NCRC),[2] which is a state institution for consumer research within the Ministry of Employment and the Economy. The NCRC was invited to join the WWViews project by the Danish Board of Technology (DBT) in March 2008, and was among the first international partners. There were three main reasons why the NCRC was inclined to accept the invitation:[3]

- the rationales of WWViews were aligned with its organizational interests (sustainability policies and climate change; enhanced participation by consumers and citizens);
- its financial situation allowed participation in a research and development project with a longer-term payback time;
- it had enthusiastic and competent staff to meet the challenge of organizing a large-scale participatory exercise.

Even though the in-house reception of WWViews was positive, it was soon apparent that other national agencies were more suspicious. Funding proved to be a critical issue. Support was sought from three different agencies, who all turned down the application.

Sitra, the Finnish Innovation Fund,[4] was the first funding agency to be consulted. A negotiation with its representative took place in spring 2008. The idea that WWViews could be funded by Sitra was rejected by one of its spokespersons who claimed that the project lacked a 'national interest point of view' required for the agency's support. Funding was then sought from the Ministry of Environment call for proposals 'to raise environmental awareness in St Petersburg, Russia'. A proposal was submitted by the NCRC in October 2008, including a plan for cooperation with a Russian WWViews partner. This time the proposal was not funded, since, according to the evaluators, the Russian policy context was not yet ready for such a participatory exercise.[5] Then funding was sought from the Nordic Council of Ministers in June 2009. The proposal was called WWViews of the Baltic Region and it included partners from Denmark, Finland, Norway and Sweden. The Council Secretariat recommended approval of the project, as it perceived the application to represent a new and interesting approach, but delegates from the Nordic countries did not support approval unanimously, which was a prerequisite. Concerns were also expressed about the date of the consultations, suggesting that it was too late to affect COP15 outcomes. The governmental position of one of the applicants (Nordregio) was also discussed. In addition to these three proposals, funding opportunities were explored in several informal discussions with other national actors but without encouraging signals. The NCRC staff finally decided not to invest more time in the applications but to carry out the project with its own budget funding.

If the funding refusals gave a negative view of the perceived relevance of WWViews among Finnish and Nordic funding agencies, more encouraging signals came from the Finnish research community. WWViews was presented by the NCRC staff in several seminars and conferences with highly positive feedback. Another indicator signalling (social) scientists' interest in WWViews is that five volunteers from three different research organizations participated as facilitators at the WWViews event.[6] NCRC staff have also been invited to lecture on university courses and seminars, and articles oriented at national researchers and policy-makers have been published (see Lammi and Rask, 2008, 2011).

A less enthusiastic approach to WWViews was expressed by Finnish politicians and policy-makers. Despite several official invitations, none of the half dozen invited policy actors (ministers, parliamentarians and high-level policy-makers) found time to officially attend the WWViews event. An even greater number of policy actors were invited to visit the WWViews event and attend the following press conference, but none accepted.

The attitude of citizens participating in the WWViews event, in contrast, was highly enthusiastic. An exit survey was distributed to all 107 participants of the WWViews event and 106 filled in the four-page questionnaire. The results were positive: 93 per cent of respondents agreed with the statement that, 'the event used my time productively'; 98 per cent of respondents were satisfied with the organization of the event; 100 per cent of participants thought that it would be beneficial to continue a dialogue process like WWViews in future.[7]

Despite several press releases and direct contacts, the media expressed little interest in WWViews. Two months after the event, the visibility of WWViews in the ten largest Finnish newspapers was nearly non-existent. Only three articles were published around 26 September, and these were based primarily on NCRC's press releases rather than presenting additional information about the hearing or its results. WWViews did not appear on Finnish TV. On Finnish web pages WWViews was mentioned roughly 40 times in blogs, news pages, e-newspapers and various kinds of information pages.

The WWViews process and its results, however, have attracted further attention from Finnish media after the conventional two-month media follow-up period (e.g. one TV interview, one radio interview and several longer newspaper articles). From the point of view of policy impact and transfer, this confirms Sabatier's (1988) remark that longer time spans are necessary for studying policy transfer; an 'action research' approach, on the other hand, that we find useful in developing new policy instruments, assumes that research into ongoing processes can contribute to reflexive development of governance processes.

To summarize, the contradictory reception of WWViews by different actors indicates a normative ambiguity related to how different actors and stakeholders – to different extents – perceive the value of the WWViews process and its results. Since impacting climate policy and strengthening (and transferring) democratic participation are among the objectives of WWViews, a more wide-ranging positive orientation by stakeholders will be necessary. Part of the normative ambiguity can be rooted in an interpretative ambiguity or unfamiliarity with the WWViews concept.[8] Different policy actors, in other words, can have quite different understandings of WWViews as a concept for deliberation: what objectives, rationales and methods it includes. While too much ambiguity can be a hindrance to the social and political support of any process, a positive aspect of ambiguity is that it allows different interpretations and helps engage different interests. Within limits, in other words, ambiguities can provide an opportunity to build more successful and widely shared models of governance (see Star and Griesemer, 1989; Heiskanen et al, 2009, p412).

Transferring deliberative democracy to local policy contexts

The transfer and impact of policy interventions as well as models of deliberative democracy have been studied from a multitude of perspectives. We build our analytical framework on two main literatures that we find most relevant in explaining the lobbying and adaptation process of WWViews and discussing alternative models of deliberation. First, we follow the approach by Heiskanen et al (2009) that builds on Sabatier's (1988) advocacy coalition framework (ACF) and related theoretical commentary (e.g. Fenger and Klok, 2001; Stone, 2004). Second, we apply a model by Renn (2008) of stakeholder and citizen participation. While the approach by Heiskanen et al (2009) provides a general heuristic model of policy impact and transfer, Renn's model provides a specific typology for studying different conceptions of deliberative democracy.

Sabatier's (1988) ACF is a widely used theory of policy processes. The core idea of the ACF model is to integrate a more traditional theoretical view of policy processes focusing on power struggles among groups with different resources and interests and operating within a given regime structure, with a focus on belief systems as causal theories by the policy actors on 'how the world operates' (Sabatier, 1988, p158). ACF focuses on policy subsystems and how 'policy oriented learning' takes place within them. In this framework policy change is seen as '... fluctuations in the dominant belief system (i.e. those incorporated into public policy) within a given policy subsystem over time' (Sabatier, 2008).

ACF theory has been elaborated and applied for different purposes (see e.g. Fenger and Klok, 2001; Stone, 2004). In this chapter we adopt the version developed by Heiskanen et al (2009) in their paper focusing on the transfer of the Dutch transition management tradition in the Finnish context of innovation policy. We not only share the (Finnish policy) context but also find their research question (the transfer of the Dutch transition management concept) parallel to ours (the transfer of the Danish WWViews concept). In our analysis of the case study we focus on the three factors that Heiskanen et al (2009, p413) find most relevant in affecting the transfer process: the receiving institutional context; the carriers of the models; the role of 'translation as an active process'.

The receiving institutional context refers to the socio-historical institutions and policy cultures that condition the policy transfer process. Transfer is, obviously, more likely between countries with similar governance systems and compatible policy cultures (Stone, 2004; Heiskanen et al, 2009, p413). The way in which the institutional characteristics of a policy arena condition governance processes has been called the 'arena effect' (Rask, 2009). For example, a participatory policy culture, such as the Danish science and technology policy, provides institutional support for the transfer of deliberative democracy, while a hierarchic policy culture, in contrast, involves resistance, compromising and reorienting effects such as a tendency to a ritualistic performance of deliberations without a real connection to policy-making (see Kothari, 2002, p149; Rask, 2009, pp64–77).

Sources and carriers of the models is the second factor in the model by Heiskanen et al (2009). Its core idea is that new governance processes are more easily adopted from certain sources. Authoritative or admired bodies, for example, are in a privileged position to spread new ideas and practices. Stone (2004) has hypothesized that non-state actors may be better at the 'soft transfer' of general policy ideas that influence public opinion and policy agendas, whereas state agencies are stronger in the 'hard transfer' of policy practices and instruments involving formal decision-making, regulation and legislation.

Both Stone (2004) and Sabatier (1988) argue that policy networks, often incorporating a tri-sectoral selection of stakeholders from the market, state and civil society, are an effective place for policy-oriented learning. If there are conflicting strategies between various coalitions, they can be mediated by a fourth group of actors, 'policy-brokers'. In Finland the position of policy-brokers in policy-making can be significant, as the bid for consensus is an important goal of creating policies (e.g. Savikko and Kauppila, 2009).

Translation as an active process refers to the idea that policy transfer is an active process involving learning, strategic and tactical playing, and active processing of the new governance concepts to make them better fit the existing institutional frame (Heiskanen et al, 2009, p414). Sometimes a new idea can be rebuffed because the recipient desires to safeguard its own mission (Powell et al, 2005). More likely, however, new ideas will be hybridized between competing models. Örtenblad (2007) has proposed that there are two different approaches for hybridization: a 'smörgåsbord' [a varied collection] strategy of carefully selecting individual elements of a complex model to fit the user's objectives, and a 'whispering game' of modifying and editing the idea throughout the transfer process (risking finally losing the original concept).

Considering the transfer (and translation) of the WWViews process we can distinguish between the different concepts of stakeholder and citizen participation considered feasible by different policy actors. In our analysis we will apply Renn's (2008) model that distinguishes between six ideal types described in Table 12.1.

In this chapter WWViews is considered to represent the anthropological concept with some features of the deliberative concept. The anthropological concept is based on the idea that common sense is the best tool to find synergies of competing knowledge and value claims. It assumes that citizens are able to make moral judgements based on their own thinking and experiences. Since common sense is a shared anthropological feature of all individuals, even a fairly small size of participating group – 12–25 people – can achieve valid results (Renn, 2008, p299). On the other hand, characteristics of WWViews that resemble the deliberative concept expect plurality of values and worldviews and aim to legitimize the decisions through consent of all parties (the recommendation stage). The communicative feature of WWViews also aligns significantly with the deliberative perspective (see Renn, 2008, pp297–299).

According to Renn (2008, pp284–295), decisions made in situations of high ambiguity, complexity and uncertainty require processes that go beyond conventional agency routines and include broad-scale deliberation and participation (see also Lidskog and Elander, 2007). International climate change negotiations are a paradigmatic example of such a decision situation. The issue then is to find the most appropriate model of deliberating[9] and expanding participation, and as the Finnish case of WWViews indicates, to embed the new model into the existing policy framework.

Considering its progressive orientation in many political questions, it may seem strange that deliberative democracy is not better known in Finland. But the question is not that straightforward, every culture must be studied in detail to understand its openness to the deliberative concept (cf Fung, 2003). As Finland is already perceived as very open politically, novel ways of exercising democracy might not have been considered necessary, or could have been viewed as threatening a system that is generally considered viable.

210 Mikko Rask and Maarit Laihonen

Table 12.1 *Six concepts of stakeholder and citizen involvement*

Concept	Main objective	Rationale	Examples of instruments
Functionalist	quality of decision output	representation of knowledge carriers; systematic integration of knowledge	Delphi, workshops, hearings, citizen advisory committees
Neoliberal	proportional representation of values and preferences	informed consent; Pareto optimality	referenda, focus groups, internet participation, negotiated rule-making
Deliberative	debating the criteria of truth and normative validity	inclusion of relevant arguments; consensus through argumentation	discourse-oriented models; citizen forums; deliberative juries
Anthropological	to engage in common sense	inclusion of disinterested lay persons representing basic social categories	consensus conference, citizen juries, planning cells
Emancipatory	to empower less privileged groups	strengthening the resources of those who suffer most	community development groups, science workshops, town meetings
Postmodern	to demonstrate variability, plurality and legitimacy of dissent	acknowledgement of plural rationalities; no closure necessary	open forums, open space conferences, panel discussions

Source: adapted from Renn (2008, p303)

The case: Understanding the WWViews adaptation process in Finland

How the policy impact and transfer of Finnish WWViews has been mediated by various factors was discussed in the theoretical section above. Contrary to the expectations of the Finnish WWViews team, media publicity was very restricted and interest from national policy actors was limited: these factors give reasons to describe the process as a short-term 'failure'.

In the following analysis we identify factors that contributed to this failure or otherwise affected the national adaptation process. Our data consists of policy-maker interviews: five pre-COP15 and three post-COP15 semi-structured interviews with Finnish COP15 delegates and other authorities working with climate policy. The interviewees represent three ministries in Finland: the Ministry of the Environment, the Ministry of Employment and Economy, and the Ministry of Foreign Affairs. A qualitative content analysis of the interviews was done to identify different ideas and concepts of deliberation, and options and challenges for citizen and stakeholder participation. Participant observations from different meetings and seminars were used as supporting data.

Socio-historical institutions: 'Arena effects' on WWViews

It has been argued that a key characteristic of Finnish policy culture is that 'everybody knows everybody': it is easy to directly contact relevant politicians and policy-makers, and impacting policies takes place through personal and social networks. Against this background, the novelty of the WWViews concept and its ideology in deliberative democracy in Finland warrants special emphasis.

The data indicate clearly that the short-term failure of WWViews around COP15 was based on long-term issues. One important factor is the Finnish understanding of civic participation as participation of civic organizations (Savikko and Kauppila, 2009). Even if final decision-making takes place in the closed circles of specific governmental and parliamentary committees, Finnish regulatory processes are typically accompanied by extensive stakeholder consultations. The preparation of the recent governmental climate and energy strategy, for example, was mentioned as a prominent example of a broad-scale consultation of societal stakeholders. A particular strength of the Finnish model, according to the interviewees, is that organizations participating in climate change politics are generally active, competent and prepared to devote their time and resources to the regulatory processes. Industrial, environmental and other interests participating in climate policies are effectively networked among themselves and between governmental bodies, and they exercise some level of coordination in their statements, which helps them coordinate their policy messages and reduce the fragmentation that would otherwise prevail.

Established actor networks are therefore the primary way to affect climate policies. Even considering the possibility of introducing WWViews as a new actor in the Finnish climate policy arena, there are still additional issues that make it a challenging case to lobby for: there is the trust in expertise and, conversely, the suspicion that citizens' participation in climate policies will be hindered by their lack of relevant knowledge; there is the conservative idea that representative democracy and its electoral voting provides an adequate channel for citizens to affect policies. Third, there is the deeply rooted consensual policy culture that does not welcome processes that focus on generating divergent opinions. When WWViews organizers attempted to gain entry to Finnish climate policy discussions, they fell into a gap: political actors were unable to identify it among traditional actors, whereas media and the masses saw it only as one survey among others.

The situation is not so monolithic, however. Even though the prevailing way of consulting stakeholder organizations was often viewed positively by the policy-makers interviewed, there were also various forms of criticism, which give reasons for rethinking existing structures of climate negotiation and identifying alternative approaches.

One weakness of the current model is, as mentioned by one policy-maker, that NGO participation is often concentrated on some vociferous key persons, which does not necessarily contribute to more democratic processes. The current model of NGOs having a role in the Finnish climate delegation was also discussed: one of the interviewees criticized this model for creating confusion between the roles and responsibilities of governmental and non-governmental bodies. This interviewee was also concerned that interest groups protect their vested interests and therefore compromise broader social goals. A more general anxiety about the current role of NGOs in international climate negotiation processes stemmed from their propensity to operate more in 'circus mode' (participating in mass demonstrations and various campaigns) than in contributory mode. It must also be noted that as NGOs do not formally represent citizens through elections, they do not have the responsibility of authoritative decisions such as those made by government officials, who must address the full range of policy issues.

What alternatives are there then for the current 'exclusive corporatist' model that could enjoy socio-historical and institutional support? Direct participation in general (e.g. referenda) was deemed by the interviewees too simplistic and inappropriate to the Finnish context. In the context of climate policy, in particular, interviewees were concerned that citizen participation would create results that undermine ambitious climate policy targets.

In addition to these sceptical notions, however, a number of insights that support the introduction of deliberative approaches – in some form or another – can be derived from the interviews. First, climate change is seen as a historical issue in the sense that citizens' support for tightening climate policies is considered necessary (some interviewees also assumed that there already is citizen support for the national climate policy), and participatory procedures were expected to contribute to broader social support. Second, there is the general appreciation of 'two-way communication' by the policy-makers, who observed that WWViews could not only provide an important channel for disseminating information about climate change, but also serve as a 'reality check' on related policies. Third, the strong tradition of futures studies in Finnish policy culture was reflected in an estimation that deliberation might have a positive role in the longer term (rather than providing solutions to short-term decision dilemmas). Fourth, ICT was mentioned as a channel of citizen consultation (in online services such as 'otakantaa.fi', where citizens can comment on ongoing regulatory processes) that could also be used as an extension of WWViews types of deliberation.

To sum up, even though WWViews encountered obstacles and indifference in the Finnish policy context, there are also factors that could help develop it into an important forum in future. Feedback from policy-makers yielded cautious but positive remarks on the feasibility of the concept, and Finnish WWViews panellists strongly supported future deliberative events. In many ways Finland can be seen as an ideal place to implement deliberative democracy, but at the same time many cultural factors can make the development very slow.

Model sources: Strengths and weaknesses

From the Finnish point of view, WWViews stands on two legs: there is the background in DBT and its tradition in parliamentary technology assessment; NCRC, on the other hand, is the national coordinator and main lobbyist for WWViews in Finland. Both facts have implications for the national adaptation process.

First, the Finnish tradition of parliamentary technology assessment (pTA) has been organized in a rather different way from the Danish one. In Finland, the key participants in pTA studies are parliamentarians themselves, while experts (and societal stakeholders) make up the other group consulted. Since limited public debate of new technologies has been identified as the main limitation of Finnish pTA activity (see e.g. Kuusi, 2004), the Danish model (oriented towards enhancing public debates) has in recent years attracted interest among Finnish pTA actors. However, the Danish type of pTA has not yet found an institutional home base in the Finnish policy framework.

Second, NCRC, which coordinated the Finnish WWViews, is a small state research institution (with a staff of some 40 persons) practising applied research in the field of consumer studies. Unlike some other research institutions, for example the similar-size Research Institute of Finnish Economy (ETLA), NCRC's role has been more distant from the preparation of governmental policies and strategies. In terms of its networks, however, NCRC is broadly linked to Finnish society: it is extensively networked with other Finnish research institutions; it cooperates routinely with policy-makers from its own Ministry (of Employment and Economy) and other ministries; and it regularly collaborates with consumer and environmental organizations.

In organizing the WWViews event, it was a conscious strategy of the NCRC team to ally with other research agencies; persons from three external research organizations actually participated as facilitators of the 26 September event. Civic organizations, on the other hand, were not invited to have a role in the organization of the citizen consultation, since the process was intended to be framed as politically neutral and scientifically organized. Two NGOs (Demos Helsinki and Dodo), however, were contacted, and the Finnish WWViews team participated in the Megapolis 2024 urban festival (organized by the latter), where it arranged a WWViews follow-up and dissemination event in central Helsinki.

When evaluating the appropriateness of NCRC to organize the WWViews event, its strength is its experience in organizing participatory research. Its status as a state-funded research agency also helps it to maintain and develop methodological competences in a systematic manner (with the help of its budget funding). It must be remembered that confidence in science, research and research institutions among the Finns is high. On the other hand, in order to adopt new deliberative and participatory approaches, Finnish policy culture may require a more gradual transfer of cultural ideas and concepts before deliberative methods can achieve their full potential (cf Sabatier, 1988; Stone, 2004). In that respect NGOs may be better positioned.

One particular implication of NCRC's status and reputation as a consumer research organization is that this can cause some confusion among media and other spectators, since the work of NCRC in the field of international climate politics can be easily understood as advocating the consumer's point of view. In fact, the consumer's point of view was strongly advocated in the Finnish recommendations (for more about this theme, see Lammi et al, Chapter 7, this volume).

The policy-makers interviewed were asked directly about who would be the best entity to organize future WWViews. Both research organizations and NGOs were considered equally appropriate. Government agencies were considered inappropriate. The underlying idea is that since deliberative methods are not part of official decision-making processes, it is natural for the organizer to be an external actor and not, for example, an official ministry. Working at the grass-roots level and in the proximity of civil society was considered a particular strength of NGOs. Whoever the organizer, extensively networking within the Finnish policy arena was considered a necessity.

Translation of citizen deliberation: Emerging ideas, competing concepts

In parallel with the complications caused by the socio-historical context and sources of the WWViews model in its adaptation, there were strong and occasionally contradictory viewpoints maintained by the policy actors concerning a feasible model of citizen deliberation.

First, the functionality and timing of WWViews was discussed by the interviewed policy-makers. Some of them thought that participatory models should not be seen as external to the process and requiring lobbying, but rather internal and firmly rooted in the decision-making process. This position may reflect the fact that the organizer (NCRC) represents research institutions traditionally understood as a reliable source of information in decision-making processes. This attitude of welcoming efforts to better integrate novel citizen deliberation processes, in fact, can be a great advantage for developing future WWViews in Finland.

Second, the question of popular representation was discussed both by the interviewees and some persons from media who interviewed the NCRC staff. Following Renn's (2008) typology (see Table 12.1), full proportional representation of the population is a major issue for the neoliberal concept (and of course, for traditional surveys), while it is not an issue for most other models, including anthropological and deliberative models that emphasize the argumentation process and qualitative aspects of deliberations. However, as long as the adequacy of 100 people to represent the Finnish population is questioned by the national actors, representation remains a thorny issue. One particular concern reflected by the NCRC team was the possibility of a green bias in the composition of the Finnish WWViews panel. A definitive answer is unlikely in this case, however, because resource limitations prevented a pre-test that might have clarified this question (see Goldschmidt et al, Chapter 5, this volume, for the German case). In spite of lingering questions, an important discussion and reflection took place around the concept of informed citizen opinion because WWViews provided a concrete example of how to create informed opinions and what they look like. Several policy-makers, NGO representatives and researchers were actually attracted to the argument that the informed opinions of the citizens that look different from the more instantaneous and less informed views created through opinion polls can potentially create space for political flexibility.

Third, the point and value of consulting citizens as lay people was understood by all policy-makers interviewed. However, there were different understandings of the appropriate amount of information and the type and specificity of questions to be asked of them. A basic philosophical dilemma is that normal citizens should be highly knowledgeable about climate policy issues to be able to respond to them, but in becoming knowledgeable, they cease to be lay people. A more practical problem is how to recruit 'disinterested lay people' (this was the requirement of the WWViews process, and it is also a characteristic of the 'anthropological model' in Renn's, 2008, typology). The Finnish experience shows that people wishing to participate in deliberations tend to be interested in the subject in question and therefore most likely to hold a predetermined opinion. Careful screening of participants is therefore necessary. The issue of appropriate level of expertise divided policy-makers' opinions. On the one hand, many doubted that the WWViews type of citizen consultations can reach a level of specificity required for relevant policy advice. On the other hand, one interviewee who was familiar with the WWViews process proposed that a more mundane version of WWViews should be developed, including simpler questions and issues which people can influence in their everyday lives.

Fourth, the role of marginal groups in deliberative consultations received extensive attention in the interviews. Many NGOs participating in climate policy negotiations represent marginal groups such as native people (Sami people in Finland), developing countries or women. Finland was described as an unofficial leader in the integration of a gender perspective in climate policy; and the integration of this aspect in policy negotiations was also mentioned as an example of civil society organizations successfully influencing government policies and even intergovernmental negotiations.

Unlike the emancipatory concept of Renn's (2008) typology, WWViews was not specifically focused on empowering any particular underprivileged group. The lack of minorities such as ethnic groups was the only point of criticism raised in the Finnish WWViews evaluation survey. Clearly it would be possible to better cover all relevant social groups in the citizen panel.

Fifth, increasing plurality of viewpoints was discussed from two different perspectives. On the one hand, the inclusion of voices from the Western world as well as perspectives of the developing world in WWViews was considered a strength of the project. On the other hand, a broader inclusion of marginal voices was considered a risk, since it leads to increased complexity in a negotiation process that is already very complex. As a solution, the interviewees suggested that the process and results of citizen consultation should be aligned with parallel channels of influence.

Conclusions and recommendations

We characterized the Finnish WWViews as a short-term failure, meaning that our expectations – before, during and immediately after the process – of how national media
and policy-makers would adopt this historical experimentation in global democracy
were not met. Based on results, however, our view of the possibility for longer-term
policy learning becomes more optimistic.

Even though it may be too early to draw strong policy lessons, we find it conducive
to more effective policy transfer to sketch alternatives on how to develop or hybridize
the WWViews concept in order to make it more easily adapted to its local context.
Based on our analysis we identify the following strategic options:

Internalizing WWViews in the national policy process

As we argued, the Finnish exclusive-corporatist culture of policy-making causes resistance to initiatives for citizen deliberation. However, there is also a strong undercurrent
of increasing understanding of its value in supporting policies. Policy-makers acknowledge that global warming is a historic issue not only in terms of its ecological dimension, but also as a challenge for renewing prevailing governance approaches. There is
a strong opinion that citizens' support for the stringent climate policies of the future is
a necessity, and understanding that WWViews types of deliberation represent two-way
communication and can provide a valuable reality check for related policies. There is
also the desire for citizen deliberation not to be seen as an external event that requires
lobbying, but rather internal and firmly rooted in decision-making processes. If future
WWViews are organized, the concept will be more familiar to policy-makers, which
means that there may be less need for lobbying and more options for coordinating the
process with other national policy processes.

Alliance with NGOs

The Finnish tradition of civic participation as participation of civic organizations should
be taken as a reality. As noted by the policy-makers interviewed, NGOs in the field of
climate policy are active, competent and prepared to devote their time and resources
to regulatory processes. They also work at the grass-roots level, which makes them
a strong partner in a citizen engagement process, not only in practical terms but also
in practising soft transfer of new governance ideas. An alliance with NGOs, however,
should not be permitted to compromise the core value that WWViews is a channel for
ordinary citizens to express their views (outside organized interest groups).[10]

Linking WWViews to action research and futures studies

The longevity of the WWViews and global citizen deliberation concept can be supported
through research and scientific institutions. The role of research is also considered a

special strength in Finland, where confidence in science, research and research institutions ranks highly. The nationally strong paradigm of futures studies provides a relevant context for methodological refinement. Framing WWViews as an action research exercise can foster its research component and pave the way for a reflexive development of governance processes.

Distinguishing informed public opinion from opinion polls

A contradiction between broad-scale citizen consultation and ambitious climate policies was discerned by some politicians, policy-makers and NGO activists. Since WWViews ended, however, with a very strong global message and requirement for highly ambitious climate policies (generally closer to the requirements of environmental NGOs than national governments), there is evidence that normal citizens, when well-informed, can have critically reflected opinions. Considering that recent incidents, such as 'Climategate',[11] severe winters in 2009–2010 and 2010–2011 across the northern hemisphere, ongoing economic crises and the failure of COP15 may temporarily reflect a more critical public opinion toward active climate policies, it can be expected, in contrast, that informed public opinion may be more immune to such events because they are based on the scientific background that has not dramatically changed in recent times. While informed opinion could be a new and untapped resource in political debates, it is also evident that it requires further scrutiny and reflection on the preconditions for using it as an element of valid policy advice.

Strengthening the local dimension of global deliberation

The promise of policy transfer is based on the idea of customizing best practice or pilot interventions to local contexts. Previously mentioned options have suggested how to mobilize local socio-historical resources for effective coordination of the local WWViews process. A more straightforward way of tuning WWViews locally could be the inclusion of local-level issues as one component of the consultation. As one interviewed policy-maker put it, not only the informative function but also the motivating function of deliberation could thus be elicited. Another way of strengthening the local dimension in national consultations would be a regionally distributed model of consultations, for example, by organizing several smaller hearings around the country instead of one bigger event.

Methodological hybridization

In a world increasingly dominated by the internet and other communication technologies, we find it striking how appealing the face-to-face component of WWViews can be: not only participating citizens but also the policy-makers interviewed expressed their appreciation of face-to-face deliberation. Hybridizing future WWViews with internet-based consultation or other options outlined above should be done with care. Much can be expected of future WWViews just by ensuring that high professional quality will be maintained under its brand name.

Notes

1 We understand lobbying as a general term referring to the efforts (by various actors) to influence policy-makers within a policy arena (see *Oxford English Dictionary*, 1989).
2 The mission of NCRC is to investigate, anticipate and identify changes and risk factors within consumer society and to communicate consumer research knowledge. Particular research specialties include sustainable consumption and participatory methods, including focus groups, participatory technology assessment and foresight (see www.ncrc.fi).
3 The authors of this chapter were among the Finnish WWViews team, which provided them with first-hand information on the situation of the NCRC during the WWViews initiative.
4 Sitra is an independent public fund under the supervision of the Finnish Parliament that promotes the welfare of Finnish society. Sitra was considered a relevant funding agency because of its role as a think tank and funding organization developing current initiatives for the strengthening of civil society and climate policy in Finland.
5 In hindsight, this assumption proved false since the Friends of the Baltic (Russian non-governmental non-profit youth environmental inter-regional organization, http://baltfriends.ru/eng_event1466?q=eng_main) successfully organized the WWViews event in Russia.
6 The partner organizations from which the facilitators were drawn included the Finnish Environment Institute (SYKE), the University of Tampere and the National Institute for Health and Welfare (THL).
7 The exit survey consisted of positive statements, to which alternative answers were selected from a seven-stage scale ranging from fully agree (1) to fully disagree (7), number (4) indicating the position of neither agreeing or disagreeing.
8 Normative ambiguity refers to the different values and preferences by different policy actors, whereas interpretative ambiguity refers to their different conceptions of the subject matter, in this case, the concept of WWViews (see Renn 2008, pp76–77 and pp151–154).
9 There is abundant literature on the advantages and disadvantages of organizing deliberative and participatory processes. While increased democracy, social acceptance and more informed decisions are the generally attributed rationales for participation (e.g. Klijn and Koppenjan, 2000; Gramberger, 2001; Bellucci et al, 2002), criticisms of these practices focus on the presumed inadequate preparation of lay people for participation in difficult policy issues; the efficiency and effectiveness of participatory processes; and the challenge of ensuring democratic representation through small-sized panels, among other issues (see e.g. Daele et al, 1997; Collins and Evans, 2002; Henkel and Stirrat, 2002; Kothari, 2002; Renn, 2004).
10 We acknowledge that this recommendation, especially, can be more relevant in Finland, where some NGOs have an influential role in decision-making processes, whereas in some other countries the situation is largely the opposite.
11 'Climategate' refers to the release of thousands of emails from the University of East Anglia's Climatic Research Unit that created controversy over the professionalism of climate scientists and the IPCC (e.g. Corcoran, 2010).

References

Aitken-Turff, F. and Jackson, N. (2006) 'A mixed motive approach to lobbying: Applying game theory to analyse the impact of co-operation and conflict on perceived lobbying success', *Journal of Public Affairs*, vol 6, no 2, pp84–101

Bellucci, S., Bütschi, D., Gloede, F., Hennen, L., Joss, S., Klüver, L., Nentwich, M., Peissl, W., Torgersen, H., van Eijndhoven, J. and van Est, R. (2002) 'Theoretical perspectives', in S. Joss and S. Bellucci (eds) *Participatory Technology Assessment: European Perspective*, Centre for the Study of Democracy, London

Bertilsson, M. (2004) 'Governance of science and technology: The case of Denmark', Discussion Paper 31, www.stage-research.net/STAGE/documents/31_Governance_of_S&T_in_Denmark_final.pdf

Collins, H. M. and Evans, R. (2002) 'The third way of science studies: Studies of expertise and experience', *Social Studies of Science,* vol 32, no 2, pp235–296

Corcoran, T. (2010) 'The cool down in climate polls', in National Post, http://network.nationalpost.com/np/blogs/fpcomment/archive/2010/01/05/the-cool-down-in-climate-polls.aspx, accessed 20 April 2011

Daele, W. v. d., Pühler, A. and Sukopp, H. (eds) (1997) *Transgenic Herbicide-Resistant Crops: A Participatory Technology Assessment. Summary Report,* Discussion Paper FS II 97 – 302, VCH Verlagsgesellschaft, Berlin, http://bibliothek.wz-berlin.de/pdf/1997/ii97-302.pdf

Fenger, M. and Klok, P.-J. (2001) 'Interdependency, beliefs, and coalition behaviour: A contribution to the advocacy coalition framework', *Policy Sciences*, vol 34, no 2, pp157–170

Fung, A. (2003) 'Associations and democracy: Between theories, hopes, and realities', *Annual Review of Sociology*, vol 29, no 1, pp515–539

Gramberger, M. R. (ed) (2001) *Citizens as Partners. Information: Consultation and Public Participation in Policy-making*, OECD, Paris

Heiskanen, E., Kivisaari, S., Lovio, R. and Mickwitz, P. (2009) 'Designed to travel? Transition management encounters environmental and innovation policy histories in Finland', *Policy Sciences*, vol 42, no 4, pp409–427

Henkel, H. and Stirrat, R. (2002) 'Participation as a spiritual duty: Empowerment as secular subjection', in B. Cooke and U. Kothari (eds) *Participation: The New Tyranny?* Zed Books, London

Järjestöjen kymmenen vaatimusta Kööpenhaminan ilmastokokoukselle (2009) www.ilmasto.org/media/koopenhamina/jarjestojen_10_vaatimusta_COP15.pdf

Kallerud, E. (2004) 'Science, technology and governance in Norway: Introduction. Science, technology and governance in Europe', Discussion Paper 19, STAGE, www.stage-research.net/STAGE/documents/19_STG_in_Norway_final.pdf

Klijn, E.-H. and Koppenjan, J. F. M. (2000) 'Interactive decision making and representative democracy: Institutional collisions and solutions', in O. van Heffen, W. J. M. Kickert and J. J. A. Thomassen (eds) *Governance in Modern Society: Effects, Change and Formation of Government Institutions*, Kluwer Academic Publishers, Dordrecht

Kothari, U. (2002) 'Power, knowledge and social control in participatory development', in B. Cooke and U. Kothari (eds) *Participation: The New Tyranny?* Zed Books, London

Kuusi, O. (2004) *Teknologian arviointitoiminta eduskunnassa. Kansainväliseen vertailuun ja kokemuksiin eduskuntakaudella 1999–2003 perustuva arvio*, Teknologian arviointeja 17, Eduskunnan kanslian julkaisu 2/2004, Eduskunnan tulevaisuusvaliokunta, Helsinki

Lammi, M. and Rask, M. (2008) 'Vuorovaikutteisella päätöksenteolla kohti globaalia vastuullisuutta', in L. Rohweder (ed) *Kasvaminen Globaaliin Vastuuseen: Yhteiskunnan Toimijoiden Puheenvuoroja*, Opetusministeriö, Helsinki

Lammi, M. and Rask, M. (2011) 'Ilmastonmuutoksen hallintaa osallistuvan päätöksenteon avulla', in L. Rohweder and A. Virtanen (eds) *Käytännön Tekoja Ilmastonmuutoksen Hallintaan*, Gaudeamus, Helsinki

Lidskog, R. and Elander, I. (2007) 'Representation, participation or deliberation? Democratic responses to the environmental challenge', *Space and Polity*, vol 11, no 1, pp75–94

Örtenblad, A. (2007) 'Senge's many faces: Problem or opportunity', *The Learning Organization*, vol 14, no 2, pp108–122

Oxford English Dictionary (1989) 2nd edn, OUP, Oxford

Pelkonen, A. (2008) 'The Finnish competition state and entrepreneurial policies in the Helsinki Region', PhD thesis, University of Helsinki, Helsinki

Powell, W., Gammal, D. and Simard, C. (2005) 'Close Encounters: The circulation and reception of managerial practices in the San Francisco Bay area nonprofit community', in B. Czarniawska and G. Sevon (eds) *Global Ideas: How Ideas, Objects and Practices Travel in the Global Economy*, Oxford University Press, Oxford.

Rantanen, M. (2009) 'Suomalaisviranomaiset eivät luota kansalaisjärjestöjen lobbaukseen', *Helsingin Sanomat*, 12 October, pA6

Rask, M. (2009) *Expansion of Expertise in the Governance of Science and Technology*, Lambert Academic Publishing, Köln

Renn, O. (2004) 'The challenge of integrating deliberation and expertise: Participation and discourse in risk management', in T. L. MacDaniels and M. J. Small (eds) *Risk Analysis and Society: An Interdisciplinary Characterization of the Field*, Cambridge University Press, Cambridge

Renn, O. (2008) *Risk Governance: Coping with Uncertainty in a Complex World*, Earthscan, London

Sabatier, P. A. (1988) 'An advocacy coalition framework of policy change and the role of policy-oriented learning therein', *Policy Sciences*, vol 21, no 2–3, pp129–168

Savikko, R. and Kauppila, J. (2009) *'Kyl se sit niissä verkostoissa tapahtuu se päätöksenteko'* – *Kansalaisvaikuttaminen ilmastopoliittiseen päätöksentekoon Suomessa* (draft: 19 October). Citizens' Global Platform: Marginal Voices Project, www.globalplatform.fi/files/kansalaisvaikuttaminen_ilmastopoliittiseen_paatoksentekoon%E2%80%A6.pdf

Star, S. L. and Griesemer, J. R. (1989) 'Institutional ecology, "translations" and boundary objects: Amateurs and professionals in Berkeley's Museum of Vertebrate Zoology, 1907–39', *Social Studies of Science*, vol 19, no 4, pp387–420

Stone, D. (2004) 'Transfer agents and global networks in the "transnationalization" of policy', *Journal of European Public Policy*, vol 11, no 3, pp545–566

TEK and UIL (2009) *Insinöörien ilmasto-ohjelma*, www.tek.fi/ci/pdf/julkaisut/insinoorien_ilmasto-ohjelma2009.pdf

13

Political Influence in the Context of Australian WWViews

Jade Herriman, Stuart White and Alison Atherton

In deliberative democracy events, good process design and a positive participant experience are important not only for the quality of citizen recommendations, but also for effective and persuasive dissemination of the results to decision-makers and the public (Rowe and Frewer, 2000). However, dissemination has received far less attention than process design in both scholarly and practitioner discourses. As deliberative practices are increasingly taken up in countries that lack the clear institutional channels and supportive political cultures enjoyed in countries with more experience in this arena, a vigorous discourse on dissemination will become an important priority. Among the questions that warrant more attention are: What does effective dissemination look like? How is it achieved? How can organizers of deliberative processes plan for effective dissemination?

The global WWViews project design required each national partner to disseminate results of their deliberation through the media and through meetings with government officials involved in developing the national negotiating position for COP15. WWViews Australia planned and delivered a multifaceted and interlinked dissemination strategy for this complex decision-making space, including direct political engagement. In this chapter we draw on this experience to understand what did and did not work well in this important component of deliberative democracy, what accounts for these outcomes, and how to do better. Our analysis incorporates both the theoretical foundations and practical considerations that informed our strategy. We also examine what other deliberative processes in Australia have revealed about the opportunities and challenges of engaging with policy-makers and change.

Our approach is to compare effort to outcomes, using both quantitative and qualitative data. For participants this includes researcher observations and field notes before, during and after WWViews day, and an exit survey administered at the end of the event. For decision-makers, we analyse key themes in verbal responses to the results, questions posed about the event, and behavioural responses to requests for participation (including sponsorship and participation in tailored briefings). A review of the literature on deliberative processes and political influence in the Australian context informs our discussion of the key challenges to dissemination. We conclude with some reflections on the broader question of how deliberative democratic processes (DDPs) are received in Australia.

Context: The experience of deliberative democratic process in Australia

The Australian political system features a federal system of government, with powers distributed between a national government (the Commonwealth) and the six states (Parliament of Australia, 2010).[1] Australia is a constitutional monarchy and parliamentary democracy; the Parliament consists of the Queen, represented by the Governor-General, and two Houses, the Senate and the House of Representatives (Parliament of Australia, 2010).

Recent research suggests that Australians are 'satisfied with and proud of the general conception of Australian democracy' (Brenton, 2008, p2), but that indications of citizen engagement are not strong, and the performance of government, politicians and other public officials are judged negatively (Brenton, 2008, p8).

The use of DDPs as a vehicle for determining citizen views is relatively new in Australia, compared to its history in the US and Europe (Carson, 2007, p1). A 2006 inventory of Australian DDPs from 1977 to 2006 documented 78 events, only 3 of which took place before the early 1990s. Nearly 40 per cent of these were convened by one portfolio of state government agencies in Western Australia under the direction of a single minister in the period 2001–2006. Almost half of the DDPs were designed, coordinated and facilitated by one person (Carson, 2007).

Past DDPs in Australia have varied both in their relationships to government and in the extent of their focus on influencing policy. Many have been primarily focused on trialling methods, or acting as pilots with the intention of demonstrating to decision-makers the potential for these methods (Herriman et al, 2007), rather than to influence or engage with a specific policy decision.

The Federal Government commissioned or funded only seven DDPs in the period of Carson's inventory, and none of these were at the national level. The five national processes inventoried were instead commissioned or funded by research or non-governmental organizations. Nonetheless, the inventory documents a slow growth of DDPs, and more have been conducted since the inventory was completed.

By far the most popular type of DDP is the Citizens Jury® (37 per cent of those documented; Carson, 2007, p3). WWViews does not fit neatly into any of the categories used in the inventory, but it included elements of a Deliberative Poll® and a consensus conference, which together account for 20 per cent of the events inventoried.

Australian DDPs scored well on representativeness and deliberativeness but low on influence in the inventory (Carson, 2007, p6). The author of the inventory therefore identifies the indifference of decision-makers and the different decision-making environments at local, state and federal levels as important research topics (Carson,

2007, p6). Evidence from subsequent deliberations, however, shows meaningful policy impact. For example, some of the local processes undertaken as part of the New South Wales Nature Conservation Council Climate Change project (NCCNSW, 2010) have seen citizen recommendations adopted by local councils in full.[2] These and other recent deliberations on climate change (Riedy et al, 2006; Green Cross, 2008; Kaufman, 2009; NCCNSW, 2009; WA Govt, 2009) show that global warming has become a high-profile issue and that Australians are very concerned about it.[3] Most of these DDPs, however, focused primarily on local (WA Govt, 2009), regional (Riedy et al, 2006), state-wide (NCCNSW, 2009) or national strategies (Kaufman, 2009). A recent DDP took up Australia's international responsibilities in the Asia Pacific region, but nonetheless lacked the global and comparative features of WWViews (Green Cross, 2008).

Political influence

Australia participated in WWViews by holding a single, national event in Sydney on 25 and 26 September 2009 with 105 randomly selected and demographically diverse Australians. The objectives of the event were to influence the Australian negotiating position at COP15 and to raise the profile of citizen dialogue processes. The Institute for Sustainable Futures (ISF) at the University of Technology, Sydney, initiated and organized the Australian event after being invited to partner by the Danish Board of Technology (DBT). The Australian sponsors included two corporate organizations, a state government department and an environmental NGO, along with the University that is home to ISF. In what follows we present the strategy of political influence that ISF devised on the basis of its previous experiences with DDPs and its reading of the Australian political context.

The importance of process design

The Australian organizers had first-hand experience in designing and delivering a range of other DDPs (White, 2001; CSIRO, 2006; Littleboy et al, 2006; Riedy et al, 2006; Herriman et al, 2007; Dryzek, 2009; NCCNSW, 2009; Office of Population Health Genomics, 2009) and were also influenced by the reflections and experiences of other Australian practitioners. In the Australian context and based on the organizers' collective experience, an important consideration was ensuring that high-ranking politicians and civil servants saw the process to be credible, legitimate and unbiased. Another consideration was enabling Australian citizens to engage meaningfully in the process regardless of their personal background or financial capacity.

These conditions – reflected through the criteria of fairness and competence[4] in evaluations of public participation methods (for example, Petts, 2001, p215) – guided most aspects of decision-making in relation to the process, where there was scope for interpreting the standardized guidelines set down by DBT (2009a). In particular, these considerations affected decisions about the recruitment of the participating citizens, as well as facilitation at the event and sponsorship.

Government-sponsored vs independent deliberation

Importantly, although sponsorship and other links were sought by organizers, WWViews was neither commissioned, organized nor sponsored by the primary actors of interest in Australia (that is, the federal government). Unlike community engagement processes initiated by the government, which ideally makes a clear prior commitment about use of the results, this case required advocacy-style communications to a variety of government recipients by the event organizers, and broad-scale media and communications efforts to raise the profile and legitimacy of the event (see Table 13.1).

Key differences in dissemination experiences of events initiated by and not initiated by government are described in Table 13.1. Given its role as an independent organizer, ISF identified the need for a proactive dissemination of the results akin to political advocacy, and looked to the partners' expertise in political advocacy to develop its approach.

The organizers' prior experience with deliberation suggested that DDPs are not well known or widely understood by Australian policy-makers. They anticipated the need to communicate what the event was, in particular its unique qualities in comparison to non-deliberative consultations such as opinion polls or public meetings with elected officials.

In the Australian context, climate change is a politicized issue that some politicians and commentators have characterized in terms of the familiar opposition of left versus right agendas.[5] Given this politicized environment, the organizers contacted all major political parties to avoid perceptions of bias or partisan alignment to create space for participants' own framing of the issues.

The components of the Australian dissemination strategy

Overall, the assessment of the local political context helped shape a dissemination strategy that included direct and targeted bipartisan political contact, targeted industry and NGO stakeholder engagement, broad media coverage featuring the experience of participants themselves, dissemination via social media, and support for participant outreach. The assessment also shaped the process design – with organizers seeking to ensure that the process was, and appeared, legitimate, transparent, unbiased and relevant to policy-makers.

The dissemination strategy consisted of four separate but interlinked strategies:

- process design supporting dissemination;
- political engagement strategy;

Table 13.1 *Risks and opportunities of government vs independent deliberations*

Link to government	Risks	Opportunities	Considerations for dissemination
Processes directly linked to government or initiated by a decision-making body	There may not be an advocate for results within the organization and results may be distributed through only one channel. The deliberation may be seen as just one of several engagement activities being undertaken (see Petts, 2001). Independence or perceived independence may be compromised.	There is a formal mechanism for considering results. Participants may have stronger confidence that results will be 'heard'. Decision-makers understand the process and approach by the time they receive results. Can be decision-making rather than consultation.	Politics within the initiating organization may result in a decision not to make the results public. There may not be an independent advocate on behalf of the citizens.
Processes not initiated by a decision-making body	Communicating the process and results can be resource-intensive. May be conflated with non-deliberative consultations, thus vying for attention in a crowded space. Difficult to identify and engage decision-makers.	Opportunity to expose policy-makers to the results, even if they are not familiar with the process. Opportunity to adopt and communicate a neutral non-partisan role in the policy process.	Dissemination of results requires a political advocacy approach. Linking the event to policy-making requires political endorsement.

- communications strategy;
- media strategy.

The following section outlines the range of activities that were undertaken as part of the dissemination strategy.[6]

Process design supporting dissemination

A top priority was to ensure that the participants were as representative as possible. To avoid the self-selection that is probably increased by recruitment methods such as advertising, invitation or snowball methods, we contracted a market research company to randomly generate telephone numbers that were used to contact prospective participants.[7] To limit the income barriers and ensure a representative mix of participants, significant costs were paid by the project, including flights to Sydney from state capital cities and accommodation in Sydney.

Another aspect of the process that was important for the legitimacy of the Australian event was facilitation. To minimize any perceptions of bias in the facilitation, organizers excluded facilitators representing organizations that may have been perceived to have a strong existing stance on the issue of climate change, and emphasized the importance of neutral facilitation in communications with facilitators, making it understood that their role was to encourage participants to express their views, not to input their own views on climate change.[8] To enhance credibility of the event, a high standard of facilitation was required – the Australian process engaged a highly experienced professional lead facilitator and recruited 33 experienced professional facilitators from sponsor organizations[9] and from professional facilitator networks.

The need for credibility and neutrality also impacted decisions about sponsorship. We actively sought to achieve a balanced sponsorship portfolio with partners from the mainstream business sector, community and government, as well as a high-profile environmental NGO.[10] We were successful in achieving this cross-sectoral mix, which helped to further boost the credibility of the event.

Political engagement strategy

The political engagement strategy for WWViews Australia had two core strands: targeted direct contact with specific decision-makers and influencers; and broad indirect influence through informing a wide network of individuals and organizations about the project, including other politicians and civil servants (at national, state and local levels), NGOs, academics and business. Dissemination objectives were considered at all stages of process design.

The organizers sought face-to-face meetings with a targeted set of individuals: government climate policy-makers and negotiators, other influential government officials, and politicians with an interest in citizen engagement. To avoid perceptions of bias, and to extend influence, the Australian government, the opposition and the Greens Party (a distinct but important player) were all directly engaged.

The organizers made early contact with key politicians and civil servants to inform them about, and engage them with, the project in advance of the event itself. As a result, the Federal Minister for Climate Change and Water, Senator Penny Wong, provided a letter endorsing the event and prepared a video message for participants; and Australia's Climate Change Ambassador, Louise Hand, spoke in person at the event.

The organizers' level of access to politicians and climate change negotiators during the period of the dissemination efforts (October and November 2009) was constrained by the limited availability of these key politicians and negotiators due to demands on their time to prepare for COP15 and the (thwarted) passage through Parliament of

Table 13.2 *Direct political engagement: Summary of effort and response*

Effort	Response	
16 politicians or negotiators approached for briefings	Number deprioritized after multiple contacts with no response	4
	Number who declined due to other priorities / commitments / generally too busy during October/November 2009	7
	Number who granted a meeting with their adviser/s or senior public servants	3
	Number of meetings with the original target	2
One party environment committee approached for a briefing meeting	Presentation granted, time slot several months after Copenhagen	1

domestic climate change legislation. Organizers attempted to arrange 16 direct political briefings of senior climate negotiators, state or federal ministers, which resulted in five face-to-face meetings (Table 13.2).

The organizers were successful in scheduling several face-to-face meetings[11] after the WWViews event, including meetings with three public servants in the Department of Climate Change (one of whom was a member of the COP15 negotiating team); the adviser on climate change to the Minister for Infrastructure, Transport, Regional Development and Local Government; an adviser to the Federal Opposition Spokesman on Emissions Trading; the Australian Greens Deputy Leader; and the Lord Mayor of Sydney, who chaired a session on citizen participation at the Copenhagen Mayors' Summit during COP15. We know that two additional senior negotiators and the Minister for Climate Change were aware of the project and its outcomes, even though meetings were not secured with them.

In addition to the targeted approach, the organizers sought to extend the impact of the project indirectly by informing as many people as possible in positions of influence about the results and process. The organizers invited politicians and other interested parties to attend the closing drinks function at the end of the event, although only a few accepted. The organizers also mailed the results report (Atherton and Herriman, 2009a) or summary report (Atherton and Herriman, 2009b) directly to all federal MPs and senators, all state government ministers, selected state government MPs, and senior federal and state civil servants engaged in climate change policy or with an interest in citizen engagement, including federal climate negotiators.

Past experiences in implementing deliberative processes demonstrated that giving participants clear messages about dissemination commitments is important for their successful participation and satisfaction with the event. This was addressed in advance information materials provided to participants (see DBT, 2009d; ISF, 2009a, 2009b). After the event, organizers provided participants with a reading list of resources about climate change (ISF, 2009d), presentation materials that they could adapt and use for giving presentations in their community (ISF, 2009g), a media release template (ISF, 2009f) and information on how to approach their local political representatives, including letter-writing tips and a letter template they could modify and send to their local politician if they chose to (ISF, 2009e).[12] The information did not advocate a particular position and the templates focused on relaying descriptive event information and group recommendations; the focus was on assisting participants who wanted to help to disseminate the results of the process or to further expand their knowledge on how to do so.

WWViews Australia was also offered space in a COP15 exhibition booth hosted jointly by the University of Technology, Sydney and the University of New South Wales, which enabled us to display copies of the WWViews Australia report and the WWViews international policy report (DBT, 2009b) at COP15.

Communications strategy

The WWViews Australia website was supplemented by a regular newsletter that went out to participants, sponsors, facilitators and other stakeholders. Seven of these newsletters were produced over the life of the project. The organizers also used these avenues to distribute a short documentary film about the Australian event that featured results as well as participant, facilitator and sponsor interviews. This was published on DVD and the project website (ISF, 2009c), and some of the content was used in a global documentary (DBT, 2009c).

To support communication about the event in participants' own communities, organizers provided participants with a template PowerPoint presentation in case they wished to make presentations to their networks – such as community groups, workplaces or even friends and family (ISF, 2009g). This action was planned prior to the event and reinforced as a priority based on participants' requests for information about what to do next. We have anecdotal evidence that four different participant-led presentations took place: a presentation to a local toastmasters group, a talk to a local Rotary Club, a community meeting to discuss the WWViews event and Copenhagen climate talks in December, and a presentation to several hundred high school students on the topic of climate change and peace.

Media strategy

Media specialists were consulted to increase television, radio and press coverage and to gain exposure on social networking sites such as Twitter and Facebook (see DBT, 2009e, 2009f). The organizers developed the key media messages that were repeated in press releases and interviews, and were successful in generating prime-time national television coverage on the day of the event, as well as national and regional radio interviews and articles in several leading state-based newspapers (e.g. ABC, 2009a; ABC Radio National, 2009; Gordon, 2009; Munro, 2009; Radio 5AA Adelaide, 2009).

Substantial regional news coverage was also achieved by focusing on the personal stories of WWViews participants (for example, *Bendigo Advertiser*, 2009; *Castlemaine Mail*, 2009; Gardiner, 2009; *Ocean Grove Echo*, 2009). The organizers also supported participants to disseminate the results in their local media and communities by providing participants with resource materials such as a media release template (ISF, 2009f).

Lessons for successful dissemination

Overall, WWViews Australia's dissemination activities performed well against the strategy – meetings with politicians and other decision-makers were held and our indirect dissemination activities exceeded expectations in terms of numbers of articles and people who were informed about the event.[13] Other measures of success included favourable participant views of the event, and their willingness to engage with media, be interviewed, and in some cases speak to other groups or local politicians about their experience. Despite this, we found engaging with the key decision-makers difficult. In the following sections, we discuss some of the key lessons from our review of the dissemination in this event.

Adequate staffing and resourcing

There is a tendency while planning WWViews-type events to emphasize the deliberation over dissemination of results, because deliberation takes place first, and its quality is crucial to having good results to disseminate. However, both are clearly important. A good deliberation that fails to reach decision-makers does not achieve its goal.

In Australia the project was carried out by a small team who were engaged in the project nearly full-time prior to the event. Time spent planning dissemination was time not spent on recruiting citizens, facilitators, organizing venue and logistics or working on the pre-event stakeholder communications. After the event there were ongoing tensions between investing time in arranging face-to-face briefings and working on the materials that were being produced to support dissemination, namely project reports and documentaries. This experience was mirrored at the international WWViews level where, for example, DBT had to find extra funding to produce and distribute an analysis of the results in a global policy report (DBT, 2009b).

Specialist skills and knowledge

'Dissemination' is the term used by both the DBT project manual and the Australian partner to describe the use of deliberative results to influence policy. This word implies the flow of information through many existing channels, and is defined as 'to spread widely' (*Oxford Dictionaries*, 2010). However, 'political engagement' or 'political advocacy' might be a more apt description of what was required after the event – both direct and indirect political engagement. This subtle shift in language captures the notion that the objectives are both political (to influence policy) and also about engagement (that policy-makers must be connected in some way to the process or results for the information to have any impact and effect change).

The organizational skill sets and experiences required to host an event that is non-partisan and neutral, both in fact and in perceptions, differ significantly from those associated with advocating for political change. The organizers found that both their NGO and business sponsors were excellent sources of information about accessing political processes, and had considerable experience and existing contacts within spheres of government. Their expertise in advocating for policy change complemented the organizers' expertise in planning and designing deliberative events that are independent and unbiased. However, the organizers chose to make initial contact with politicians themselves, which reinforced the independence and neutrality of the process.

We also engaged specialist media and communications assistance to increase media coverage and invest in intensive stakeholder communications. Again, this expertise may not always reside in the organizations collaborating on a deliberative event. Partnerships with advocacy organizations, coalitions of NGOs or specialist media and public relations experts could be valuable for future deliberative events.

An early start to dissemination activities

Dissemination should be conceived as a whole-of-project endeavour and not just something that happens once the results are out, although clearly the focus on dissemination intensifies once the results are available. Our experience confirmed that there are many valuable opportunities to raise awareness of a process like WWViews in the planning stages, long before the final outcomes are known, as process design, decisions about sponsors, and stakeholder communication were all developed with a clear objective of maximizing dissemination potential alongside more immediate goals. In addition to strengthening dissemination itself, this approach informed other crucial decisions about the project. For example, the importance of demographically representative participant groups, neutral and skilled facilitation and carefully considered sponsor relationships is magnified when viewed through the lens of dissemination, because the credibility of the deliberation becomes a central issue when delivering the results to external parties. This underlines the importance of linking dissemination objectives with process design at the outset of the project.

Dissemination is strengthened when key stakeholders have heard of the event and have a sense of its process before seeing the outcomes, and our experience shows that face-to-face meetings are the best way to achieve the depth of connection required. This is especially important when dissemination aims to educate policy-makers on the process and its potential as well its outcomes. Next time, we would start earlier with pre-event briefings in order to ensure that the target audience has some advance familiarity with the event that strengthens their interest in receiving results, and in order to secure early commitments to post-event follow-ups.

The WWViews events were held in late September, and results were available immediately, but the Australian report and the international report were not available until mid-November for the COP15 summit that commenced in early December. This left only a short time for direct communication of results to policy-makers, and in all likelihood missed key timelines for affecting final policy positions.

An opportunistic and adaptive dissemination approach

One political briefing with New South Wales State Member of Parliament, the Honourable Clover Moore (who is also Lord Mayor of Sydney where WWViews Australia was held) was especially productive for domestic and international dissemination. After meeting with organizers, Ms Moore communicated the WWViews results to all constituents in her municipality; used the DVD in public meetings about climate change; and expressed her interest in amplifying the WWViews results in Copenhagen, where she already planned to attend the Mayoral Summit held in conjunction with COP15. Through liaising with DBT, we were able to secure an invitation for Ms Moore to present the WWViews global results to other civic leaders in a special address to the Mayoral Summit.

This experience demonstrates the potential communication and dissemination benefits to be gained from aligning dissemination objectives with the political commitments and objectives of targeted decision-makers. This requires a flexible approach to allocating resources to dissemination follow-up, and may require creating and providing tailored documents or resources to meet their needs.

We also have anecdotal evidence that participants spoke to other groups or local politicians after the event, and we supported this by creating presentation materials that were posted on the project website (ISF, 2009g). A challenge for organizers was creating this material fast enough to meet the interest of participants directly after the event, while also managing report writing, sponsor liaison and other urgent post-event tasks. Unfortunately, the project team lacked the resources for monitoring and evaluation of these participant-initiated communication and lobbying activities. An initiative for future WWViews-type events might be tracking and further supporting participant-directed follow-up with policy-makers, and perhaps using this information in meetings with other politicians. Raising the number and types of participant meetings with politicians could provide convincing evidence of grass-roots support for the deliberative results and demonstrate citizen interest in being heard by decision-makers, especially if this information were to be consolidated and disseminated at the global level.

Direct engagement with policy-makers

In Australia, direct access to politicians is possible but restricted. The process of securing a meeting with a politician or government official required time and persistence and was difficult when there was no existing relationship. A specific constraint on access to politicians and climate change negotiators for WWViews was their limited availability due to preparation for COP15 and the passage through Parliament of domestic climate change policy.

The response rate to our efforts to engage politicians (5 meetings of 16 requested) seemed low in light of the multiple contacts made via multiple media over a two- to three-month period, the media and citizen interest in climate change over this period, and the fact that WWViews Australia was linked to a global decision-making process. The particular period of time for these briefings coincided with climate negotiations in Bangkok and Madrid. This was a significant competing interest (and higher priority) for politicians or negotiators on our list, who were absorbed with preparations for and attendance at these events. However, it should also be noted that at least four of the seven who firmly declined our invitation gave no specific reason for being unavailable beyond being generally busy, and showed no willingness to discuss possible dates or times. For these, the organizers infer that the project or the results were not of sufficient interest to warrant a meeting, or are seen as a low priority compared to other issues and commitments.

The response from politicians we met, by contrast, seemed to indicate an increased interest in the concepts underlying the project and increased awareness of the results. Face-to-face meetings resulted in follow-up activity (one possible proxy for interest and engagement) in a way that emails, letters and phone calls did not. We therefore conclude that these face-to-face meetings contributed significantly to the objectives of the dissemination strategy. This highlights the importance of sustained engagement and a two-way dialogue, as opposed to a passive distribution of results in a mass mailing or through mass media.

In future the organizers would invest more effort in direct briefings and even run pre-event briefings. The value of early stakeholder familiarity with the project is especially important when the dissemination aims to educate policy-makers about the process and its potential as well as the outcomes. However, ultimately such efforts are at the mercy of changing circumstances and diaries.

Understanding the culture of decision-making in Australia

Where a deliberative process is not commissioned by a decision-making body, as was the case with WWViews Australia, the likelihood of achieving its ultimate aim to influence decision-making can depend on many contextual aspects largely beyond the control of the event organizers, including how such processes are viewed by policy-makers (Petts, 2001). As noted earlier, policy-makers' unfamiliarity with DDPs created additional challenges for the political engagement strategy.

The Australian WWViews event could be considered a hybrid in terms of its relationship to government – neither initiated and commissioned by government nor completely independent of government. Organizers sought to create a sense of national government support for the event that would raise its profile and sense of legitimacy through early outreach to government officials seeking speakers, endorsements and sponsorship in addition to briefings. We emphasized international alignment by noting that COP15 host Connie Hedegaard was a WWViews Global Ambassador, but this had no discernable effect.

During initial discussions about sponsorship, one senior government official explained that 'we already have our schools' program for climate change' and 'we invited submissions on the (climate change) green paper'. While this comment may simply indicate that available resources were already expended for other engagement activities, it might also betray a conflation of DDPs with community education and unstructured input. Either way, early contact with decision-makers that effectively conveys the unique qualities of DDPs should clearly be a priority for organizers.

In the briefings that ultimately were conducted, politicians and top government officials were interested and engaged. They posed incisive questions, demonstrated gratitude for the organizers' efforts to share the results and commented favourably on the project. In some meetings there was an interest in the results of specific countries that were considered important to the negotiations, showing the attendees' detailed engagement with the results. The details of participant recruitment techniques were a key point of discussion in at least two meetings, as were comparisons of results in Australia with other countries. Participants in several meetings asked probing questions about future WWViews processes, and in four of the five meetings staff enquired about ways in which the results had been used, who had seen the information and what their response had been.

Attendees at these sessions were unsurprisingly positive about results that supported existing policy directions, or that confirmed prior polling on public interest in the issue (see note 3). Their questions revealed attention to and an interest in specific issues, especially recruitment methods. In addition to this engagement with the results, four of the five meetings produced offers of follow-up action to promote the results or make others aware of the project.[14] Notwithstanding the active engagement with WWViews in these briefings, however, the nature of the detailed questions and the occasional use of language such as 'conference' to describe the event revealed an incomplete grasp of DDPs versus other group forums. This mirrored our experience with officials prior to the event.

Petts (2001, p221) suggests that deliberative processes should be concerned not only with immediate impacts, but also 'the broader, perhaps longer-term, culture of decision making. Deliberative processes should make a significant contribution to changing the culture in which decisions are made.' Petts also notes that local authorities who have been involved in DDPs have begun to develop new relationships with the public. 'There is recognition that decisions can be better informed. The confidence of decision makers to take difficult and complex decisions has been raised' (Petts, 2001, p222). This supports the aim of WWViews to demonstrate and promote the role of deliberative processes in decision-making. Deliberative inclusive processes are counter-cultural to the conventional approaches to political decision-making in Australia, despite the important exceptions described by Carson (2007). This novel and marginal status is evidenced for WWViews by the difficulty that the organizers encountered in securing interested audiences for the results. Still, confidence in the ability of our current political system and processes to solve the complex and 'wicked' problems of our age is clearly diminishing (Rittell and Webber, 1973, p159; Carson and Martin, 1999, p8). This failure of the conventional decision-making system may provide fertile ground for the growth, and ultimately increasing legitimacy, of deliberative inclusive processes (White, 2008, p6).

Conclusions

Political influence is unpredictable and hard to achieve, especially for events not formally part of the decision-making process. Influencing outcomes cannot be guaranteed. This prompted us to apply a diversity of approaches to maximizing the success of the political engagement process in a complex landscape. To do this we created a set of interlaced strategic approaches. We think we were successful in achieving a broad range of direct and indirect contacts with decision-makers, some of which seemed to have positive results, although the effectiveness of direct contact was limited by the time it required and the uncertain outcomes we perceived.

The political influence of any process will be tempered by the local political and cultural landscape. In this case, the Australian backdrop includes low levels of influence to date from DDPs, and generally limited understanding of the process and its features. This implies that it will be harder for DDP organizers to get meetings, sponsorship or endorsement than is the case for well-organized lobbyists and advocacy groups. The favourable response in face-to-face meetings probably reflects the improved understanding that these encounters afford, which is a clear sign that outreach to decision-makers has to be grounded in learning about deliberative processes as well as the policy issue under consideration. In contrast to the challenge that the novelty of WWViews posed for engaging decision-makers, citizens and facilitators viewed this feature as an attraction, expressing interest in being involved with the process as well as the topic itself.

WWViews-type events that sit outside normal decision-making frameworks and are not commissioned by decision-making authorities require dissemination to be a core part of the work, receiving significant thought, planning and resourcing. Without this, the unique voice that DDPs aim to construct can become just another voice – and a weak one. To be credible, these projects need to avoid the confrontational political strategies of advocacy campaigns, but this means that they lack the broad public appeal of well-designed media events. Fortunately, avoiding partisan engagement and advocacy does not preclude the judicious use of techniques familiar in advocacy and campaigning. The question is whether organizations skilled in running deliberative processes are likely to be best placed for performing this function. The experience of the global WWViews process on this question is mixed. We further note the temptation in any deliberative process to focus almost entirely on the deliberation itself and overlook the need for ongoing effort to ensure that its outputs are utilized. We certainly found this to be a challenge in our WWViews experience.

Overcoming the marginal status of DDPs requires more effective dissemination. The response of those decision-makers that we briefed indicates that face-to-face meetings and other direct contact must be a central strategy in dissemination processes. Organizers will need to exercise patience, persistence and realistic expectations over time to familiarize policy-makers with DDPs and establish a counterpoint to the modes of citizen participation they currently use. To support this, not only increased resources, but also better attention to political engagement strategies before, during and after the processes, will be essential.

Acknowledgements

The authors gratefully acknowledge the editors of this volume and Dr Kath Fisher and Nicole Thornton for their advance review and thoughtful comments on this chapter. They would also like to acknowledge the other members of the WWViews Australia project team: Amber Colhoun, Jennifer Croes, Dr Chris Riedy and Rebecca Short; and thank all the organizations and individuals who helped to make WWViews possible. In particular we are grateful to: the partners in the international WWViews Alliance, especially the Danish Board of Technology and the Danish Cultural Institute; the WWViews Australia sponsors (University of Technology Sydney, PriceWaterhouseCoopers, National Australia Bank, WWF Australia and the Department of Sustainability and Environment Victoria) and other supporting individuals and organizations; the volunteer facilitators and event logistics team; and the WWViews participants themselves.

Notes

1 Three territories – the Australian Capital Territory, the Northern Territory, and Norfolk Island have self-government arrangements.

2 Personal communication (8 June 2010) from Dr Kath Fisher, Southern Cross University, an organizer and facilitator in the New South Wales project.

3 For example, recent surveys reported that an overwhelming majority of Australians (93 per cent) have heard of global warming (ACNielsen, 2007) and believe global warming is occurring (84 per cent – Carson et al, 2008; Newspoll, 2008). In other polls, a large majority were concerned about climate change (77 per cent: The Climate Institute, 2009; and 73 per cent: ABS, 2009), and an overwhelming majority (93 per cent) believed that climate change and its

effects pose a problem for Australia. Only 5 per cent of respondents believed climate change was not a problem for Australia (Newspoll, 2007).

4   These criteria focus on the deliberative process rather than its influence on decision-making. For example, Timotijevic and Raats (2007, p309–310) propose outcome measures for evaluations that focus solely on the participants' experience of the event. Van Kasteren and McKenna (2006, p24) argue that a positive disposition by participants 'towards the process … enhances the likelihood of CJ [citizen jury]-based policies being adopted by the council and the company', but they provide no direct measures of policy outcomes.

5   For example, a senior Liberal Party politician has publicly accused the left of exploiting people's innate fears about global warming and climate change to achieve their political ends (ABC, 2009b) and using it as an opportunity to 'de-industrialize the western world' (ABC, 2009b).

6   Many dissemination ideas were considered in addition to those implemented. For example: engaging a WWViews Australia ambassador, targeting marginal political seats, public addresses at conferences and other forums, hosting larger-scale political briefings, organizing a high-profile launch or press conference for the Australian report, and sending a delegation, including participants, to Copenhagen to take part in international dissemination efforts during COP15.

7   For more information about recruitment methods and the resulting mix of participants, see Atherton and Herriman, 2009a.

8   This was emphasized in the detailed information packs, facilitator agreement forms and pre-event training materials.

9   With the exception of WWF, an environmental NGO that was excluded from facilitating to avoid any perception of bias.

10 The sponsors were University of Technology Sydney, PricewaterhouseCoopers, National Australia Bank, WWF Australia and the Department of Sustainability and Environment Victoria.

11 These briefings ranged from 30 to 90 minutes in length and generally included a formal presentation followed by questions and answers. Two meetings were attended by one politician/representative; the others were attended by three or more people. Each briefing included the distribution of the project report and summary reports, and viewing the five-minute project documentary. Other print materials included copies of media articles and a table of Australian results compared to world averages.

12 Available at www.wwviews.org.au/participants/whatnow.

13 Recipients of the results or summary reports included all federal politicians (MPs and senators), all state government ministers, selected state government MPs, and senior Federal and State civil servants engaged in climate change policy or with an interest in citizen engagement, including federal climate negotiators; a project newsletter was distributed to several hundred stakeholders on seven occasions over the life of the project.

14 Actions proposed in discussions included additional briefings or reports for candidates suggested by attendees. Some (three out of five) volunteered to link to the results from their website and one of these also volunteered to refer to materials in forthcoming speeches and community forums. One meeting resulted in an offer to make email contact with other key people within the Department to mention the briefing meeting and project; another meeting resulted in an offer to pass on briefing materials to an adviser to another politician within the same party.

References

ABC (2009a) 'ABC TV 7pm news report', 26 September 2009, Australian Broadcasting Corporation, Sydney, www.youtube.com/watch?v=JFI7xZQNqz8, accessed 21 December 2009

ABC (2009b) 'Malcolm and the malcontents', Australian Broadcasting Corporation, Sydney, Transcript of Sarah Ferguson's report on ABC's *Four Corners* program, first broadcast 9 November 2009, www.abc.net.au/4corners/content/2009/s2735044.htm

ABC Radio National (2009) 'World Wide Views on Global Warming', *Future Tense*, ABC Radio National, 22 October, www.abc.net.au/rn/futuretense/stories/2009/2714757.htm, accessed 21 December 2009

ABS (2009) '4626.0.55.001 – Environmental views and behaviour, 2007–08 (2nd issue): Summary of findings', Australian Bureau of Statistics, www.abs.gov.au/AUSSTATS/abs@.nsf/Latestproducts/4626.0.55.001Main%20Features22007–08%20%282nd%20issu e%29?opendocument&tabname=Summary&prodno=4626.0.55.001&issue=2007–08%20 %282nd%20issue%29&num=&view

ACNielsen (2007) 'Global warming: Are you aware of it? Is it self inflicted? Is it a serious problem? According to the Australian population, the answer is yes', News Release, http://au.nielsen.com/news/documents/GlobalWarmingreleaseFeb2.pdf, accessed 25 August 2009

Atherton, A. and Herriman, J. (2009a) *The World Wide Views on Global Warming Australia Story*, Institute for Sustainable Futures, University of Technology, Sydney

Atherton, A. and Herriman, J. (2009b) *The World Wide Views on Global Warming Australia Story Summary*, Institute for Sustainable Futures, University of Technology, Sydney

Bendigo Advertiser (2009) 'Climate change a hot issue', 30 September, http://wwviews.org.au/uploads/bendigoadvertiser.pdf, accessed 21 December 2009

Brenton, S. (2008) 'Public confidence in Australian democracy', Democratic Audit Discussion Paper 8/08, December, http://democraticaudit.org.au/?p=55

Carson, L. (2007) 'An inventory of democratic deliberative processes in Australia: Early finding', January 2007, www.activedemocracy.net/articles/engaging%20comm%20summary%20 070115.pdf

Carson, L. and Martin, B. (1999) *Random Selection in Politics*, Praeger Publishing Westport, CT

Carson, R. L., Louviere, J. J. and Wei, E. (2008) 'Structuring Australia's Climate Change Plan: The public's views', Draft 26 August, University of Technology, Sydney, www.censoc.uts.edu.au/pdfs/working_papers/wp08002.pdf

Castlemaine Mail (2009) 'Local input on world problem', 2 October, http://wwviews.org.au/uploads/castlemainefull.pdf, accessed 21 December

The Climate Institute (2009) *Climate Institute Fact Sheet: Tracking Climate Change Attitudes (May)*, www.climateinstitute.org.au/images/polling_report_may09.pdf, accessed 22 August 2009

CSIRO (2006) 'The heat is on – the future of energy in Australia', A report by the Energy Futures Forum, Commonwealth Scientific and Industrial Research Organization, ACT, Australia, December, www.csiro.au/files/files/pbew.pdf

DBT (2009a) *World Wide Views on Global Warming – A Global Citizen Consultation on Climate Policy* (Process Manual), The Danish Board of Technology, Copenhagen

DBT (2009b) 'World Wide Views on Global Warming – From the world's citizens to the climate policy-makers: Policy report', The Danish Board of Technology, Copenhagen, http://

wwviews.org/files/AUDIO/WWViews%20Policy%20Report%20FINAL%20-%20Web%20 version.pdf, accessed 21 December 2009

DBT (2009c) 'World Wide Views on Global Warming – The documentary', The Danish Board of Technology, Copenhagen, www.wwviews.org/node/236, accessed 21 December 2009

DBT (2009d) 'World Wide Views background reading', The Danish Board of Technology, Copenhagen, http://wwviews.org.au/uploads/wwviews%20climate%20change%20 background%20reading.pdf, accessed 21 December 2009

DBT (2009e) 'World Wide Views blog', The Danish Board of Technology, Copenhagen, http:// blog.wwviews.org, accessed 21 December 2009

DBT (2009f) 'World Wide Views on Global Warming – Facebook page', The Danish Board of Technology, Copenhagen, www.facebook.com/pages/World-Wide-Views-on-Global-Warming/76775305887?ref=mf, accessed 21 December 2009

Dryzek, J. (2009) 'The Australian citizens parliament: A world first', *Journal of Public Deliberation*, vol 5, no 1, article 9, http://services.bepress.com/jpd/vol5/iss1/art9, accessed 21 December 2009

Fisher, K. (2010) Personal communication about the experience of the Nature Conservation Council of NSW Climate Consensus project and specifically the local forums held in partnership with NSW local councils, February

Gardiner, P. (2009) 'Peter answers the call for a change', *Noosa News*, 9 October 2009, www.noosanews.com.au/story/2009/10/09/peter-answers-the-call-for-a-change, accessed 21 December 2009

Gordon, J. (2009) 'Carbon scheme in the bag', *The Age*, 15 November 2009, Melbourne, www. theage.com.au/environment/carbon-scheme-in-the-bag-20091115-ifwc.html, accessed 21 December 2009

Green Cross (2008) 'A national people's assembly: Addressing the humanitarian consequences of climate-induced sea level rise in Asia Pacific, and Australia's response – Recommendations to the Australian Federal Government', 29–30 August, Green Cross Australia, Brisbane, www. greencrossaustralia.org/media/103898/national_peoples_assembly_final.pdf, accessed 21 December 2009

Herriman, J., Plant, R. and Chong, J. (2007) *Yarra River Values Forum – A Citizens' Forum held in Ivanhoe*, 1–3 December 2006, Volumes 1 & 2, Institute for Sustainable Futures, Sydney

ISF (2009a) *World Wide Views on Global Warming Information Pack*, 10 August 2009, unpublished, available by request from Institute for Sustainable Futures, University of Technology, Sydney

ISF (2009b) *Participant Agreement Form*, 7 August 2009, unpublished, available by request from the Institute for Sustainable Futures, University of Technology, Sydney

ISF (2009c) *World Wide Views on Global Warming Australia – Mini-documentary*, 9 November 2009, www.youtube.com/watch?v=UvHca7kQySo, accessed 21 December 2009

ISF (2009d) 'Post-event information for participants', Institute for Sustainable Futures, University of Technology, Sydney, http://wwviews.org.au/uploads/post%20event%20info%20for%20 participants.pdf, accessed 21 December 2009

ISF (2009e) 'World Wide Views on Global Warming – Letter template', Institute for Sustainable Futures, University of Technology, Sydney, http://wwviews.org.au/uploads/ worldwideviewsonglobalwarminglettertemplate.doc, accessed 21 December 2009

ISF (2009f) 'World Wide Views on Global Warming – Media release template', Institute for Sustainable Futures, University of Technology, Sydney, http://wwviews.org.au/uploads/mediarelease_participant_in_global_climatechangetalks.doc, accessed 21 December 2009

ISF (2009g) 'World Wide Views on Global Warming – Participants' presentation', Institute for Sustainable Futures, University of Technology, Sydney, http://wwviews.org.au/uploads/presentationforparticipants.ppt, accessed 21 December 2009

Kaufman, S. (2009) 'Rising above hot air: An experiment in deliberative dialogue on a major collective dilemma', Paper presented at *Greenhouse 2009*: Climate Change and Resources, Perth, 23–26 March 2009, www.greenhouse2009.com/downloads/Communicating_090325_1100_Kaufman.pdf, accessed 21 December 2009

Littleboy, A., Niemeyer, S., Fisher, K. and Boughen, N. (2006) *Societal Uptake of Alternative Energy Futures: Final Report*, CSIRO Exploration and Mining Report No P2006/784, Brisbane

Munro, K. (2009) 'Australians back climate action', *Sydney Morning Herald*, 28 September, www.smh.com.au/environment/australians-back-climate-action-20090927-g7qq.html, accessed 21 December 2009

NCCNSW (2009) *NSW Community Climate Summit 19–21 February 2009: Recommendations to the NSW Government*, March 2009, Nature Conservation Council of NSW, Sydney, www.nccnsw.org.au/images/stories/nsw_climate_summit/summit_report.pdf, accessed 21 December 2009

NCCNSW (2010) 'NSW Climate Summit – more about the project', http://nccnsw.org.au/index.php?option=com_content&task=view&id=2488&Itemid=1133, accessed 13 February 2010

Newspoll (2007) 'Climate change poll (21/02/07)', http://newspoll.com.au/image_uploads/0205%20Climate%20Change.pdf, accessed 26 August 2009

Newspoll (2008) 'Climate change (29/07/08)', http://newspoll.com.au/image_uploads/0708%20Climate%20Change%2029–07–08.pdf, accessed 25 August 2009

Ocean Grove Echo (2009) 'Exchanging climate views', *Ocean Grove Echo*, 8 October, http://wwviews.org.au/uploads/oceangroveecho.pdf, accessed 21 December 2009

Office of Population Health Genomics (2009) *Guidelines for Human Biobanks, Genetic Research Databases and Associated Data*, Perth, Department of Health, Government of Western Australia

Oxford Dictionaries (2010). Oxford Dictionary of English, third edition. Angus Stevenson (ed.). Oxford University Press, Oxford, UK.

Parliament of Australia (2010) 'Parliament: An overview', www.aph.gov.au/parl.htm, accessed 4 February 2010

Petts, J. (2001) 'Evaluating the effectiveness of deliberative processes: Waste management case-studies', *Journal of Environmental Planning and Management*, vol 44, no 2, pp207–226

Radio 5AA Adelaide (2009) 25 September 2009, dmgRadio Australia, Adelaide

Riedy, C., Atherton, A. and Lewis, J. (2006) 'Capital Region Climate Change Forum: Citizens' report', prepared for the NSW Greenhouse Office and ACT Department of Territory and Municipal Services by the Institute for Sustainable Futures, December, www.isf.uts.edu.au/publications/riedyetal2006citizensreport.pdf

Rittell, H. W. J. and Webber, M. M. (1973) 'Dilemmas in a general theory of planning', *Policy Sciences*, vol 4, no 2, pp155–169

Rowe, G. and Frewer, L. (2000) 'Public participation methods: A framework for evaluation', *Science, Technology & Human Values*, vol 25, no 1, pp3–29

Timotijevic, L. and Raats, M. (2007) 'Evaluation of two methods of deliberative participation of older people in food-policy development', *Health Policy*, vol 82, no 3, pp302–319

van Kasteren, Y. and McKenna, B. (2006) 'The micro processes of citizen jury deliberation: Implications for deliberative democracy', *Proceedings of the 56th Annual Conference of the International Communication Association*, Dresden International Congress Centre, Dresden

WA Govt (2009) 'Climate change: Sustainability within a generation', Government of Western Australia, Perth, http://home.getinvolvedwa.org/home.html, accessed 21 December 2009

White, S. (2001) *Independent Review of Container Deposit Legislation in New South Wales: Final Report – Volume III*, Prepared for the Hon. Bob Debus MP, Minister for the Environment by the Institute for Sustainable Futures, November, www.isf.uts.edu.au/publications/whiteetal2001depositsNSWvol3.pdf, accessed 21 December 2009

White, S. (2008) 'Pathways to deliberative decision-making: Urban infrastructure and democracy', paper presented to Conference on Environmental Governance and Democracy, Yale University, New Haven

14

Seeking the Spotlight: WWViews and the US Media Context

Jen Schneider and Jason Delborne

This chapter focuses on the development and implementation of various media plans and strategies for WWViews in the US. While we aim to consider the US case within the larger context of global media coverage of WWViews, we focus primarily on the particular coverage the US received, both because this is the context with which we are most familiar, and because the lessons learned in the US may apply across multiple contexts, given the domination of globalized media conglomerates worldwide (though allowances must be made for the tremendous diversity of political and social contexts around the world). We therefore attempt to understand the successes and shortcomings of WWViews media strategies in the US and to offer thoughts on how future deliberative projects might garner media coverage more effectively.

In further pursuit of a constructive critique, we also pose some questions about the purpose of getting media coverage for a project like WWViews and the problems that may arise when organizers rely upon media coverage as a tool for policy dissemination. In particular, our analysis suggests that media coverage should not be a primary goal of events such as WWViews: rather, media should be used as a form of citizen outreach, a goal more in keeping with the vision of the WWViews emphasis on citizen deliberation. Focusing predominantly on media coverage, by contrast, can have unforeseen and unwelcome effects.

Background

WWViews planners at the national and international levels paid considerable attention to media and dissemination as goals of the WWViews process. WWViews planners in Denmark – staff of the Danish Board of Technology (DBT) – requested media plans from all national partners; and national partners at the five American sites (Boston, MA; Atlanta, GA; Denver, CO; Phoenix, AZ; and Los Angeles, CA) all made efforts to secure media attention for their local events. A review of planning documents suggests that, from the perspective of most organizers, WWViews should have been an obvious media attraction: there had never been a global citizen deliberation event like this; there was extraordinary hype leading up to the COP15 talks; and citizens were making their voice heard in a new kind of deliberative forum that spanned five continents. This was democracy in action, and certainly something worthy of media coverage.

Yet, our research on WWViews indicates that in the US there was little to no media coverage before the event, some notable but spotty local and national coverage in the weeks immediately following the event, and then a significant drop-off in coverage. While we recognize the uncertainty of achieving media coverage with any strategic plan, our analysis provides some insight for understanding the disconnect between hopes and outcomes.

Methods

Telling the story of WWViews is challenging, in part because there were over 40 participating sites worldwide, each with a different social, political and cultural context. Furthermore, US researchers alone have amassed an enormous amount of data, ranging from dozens of ethnographic interviews to hundreds of surveys and questionnaires. Even our narrow focus on media presents significant challenges: to date, an exhaustive list of media coverage across the globe has not been possible to collect; and much of the US coverage has occurred on the internet, creating a somewhat confusing record of the proliferation or replication of blog posts and similar opinion editorials as they are posted and reposted across multiple sites.

In light of these limitations, the purpose of this chapter is not to provide an exhaustive list of media coverage within the US and worldwide (if that were possible), nor do we attempt to perform a complete content analysis of major coverage. Rather, our goal is to tell the story of media outreach for the WWViews project in the US and examine the assumptions we and other project planners made about media coverage and its value. As a result, we rely on personal emails and correspondence with other WWViews planners and organizers; notes from the many conference calls the US team participated in during 2009 and 2010; internal newsletters and planning and funding documents distributed by DBT and US organizers; media and dissemination plans from the various US sites, DBT and some international sites, mostly in northern Europe; media artefacts themselves, including articles, blog posts, radio shows and television appearances; interviews with site and media coordinators; and our own observations and experiences at one WWViews site (Denver) and at the COP15 meetings. In sum, we saw ourselves – in very modest terms – as participant observers (Burawoy et al, 1991; Frey et al, 2000; Lindlof and Taylor, 2002; Denzin and Lincoln, 2008).

We refrain from making claims about the effect the project's media coverage had on policy outcomes, because such causality would be fragile at best. We acknowledge that we may have overlooked some instances of coverage in the US and internationally, but provide evidence to support our claim that, given the media landscape as a whole, WWViews did not penetrate mass media enough to have prompted or significantly nudged particular policy outcomes. This is not to suggest that media coverage in and of itself is not worthy of pursuit for a project like WWViews. As a form of educational or public outreach, it might be quite useful and important. In fact, in the case of the Boston Museum of Science, outreach was a successfully met goal. However, we suggest caution in placing media coverage as a central strategy to influence policy outcomes.

Programmatic commitment to WWViews media strategies

Our data make clear that international and national event planners had media engagement and dissemination in mind from the early planning stages. At the highest level – among the DBT staff who conceived WWViews as a project and coordinated the global effort – media was central to all stages of the process. Our interviews with DBT organizers revealed that, while limited by staff and monetary resources, the WWViews project manager, executive director and several interns expended enormous energy in seeking media coverage in Denmark (the host country of the COP15 meetings) and around the world. Specifically, they produced a media-friendly policy report that distilled the results of the WWViews consultation (DBT, 2009b); invited ambassadors and press to a meeting at the Danish Parliament on 19 November 2009; organized an official side event at the COP15 meetings; held a similar event at Klimaforum, the well-attended alternative conference in Copenhagen during COP15, primarily designed for those NGOs and individuals who lacked access to the official COP15 meetings in the Bella Center; and worked successfully with a Danish TV show to produce a significant piece on WWViews during a time slot typically viewed by 20 per cent of the Danish population. In their role as global project coordinator, DBT also created a forum on the WWViews intranet for media strategy; sent out draft press releases to national partners on several occasions; and requested media plans from all national partners.

Balanced with this enthusiasm, however, DBT staff acknowledged in interviews two programmatic shortcomings in the WWViews media strategy. First, DBT failed to attract an institutional partner with significant access to global media (e.g. major foundations, official United Nations bodies or major international media such as the BBC). According to DBT staff, this meant that, for all practical purposes, there was 'no global media plan'. Second, DBT faces an ongoing struggle to position its work as neutral and non-partisan. Such positioning dampens their ability to make strong claims about issues that would contribute more to an aggressive media plan but resonate more typically with advocacy organizations. DBT staff members acknowledge this tension forthrightly in their work for the Danish government in conducting technology assessments, and in particular with respect to the WWViews project.

Nevertheless, early planning documents demonstrate that WWViews planners saw media outreach as an important part of the dissemination goals of the project. Nearly all early planning and funding documents from DBT and in the US contained the following language: 'The WWViews deliberations will be held worldwide during a single 36-hour period, and publicized immediately via the World Wide Web, building excitement, drama, and media interest throughout the day' (US National Planners, 2009). In an Impact Strategy document circulated in 2008, planners argued that, 'Media coverage will bring project results to the attention of everyday people, interest groups, and political decision makers worldwide. These people may, in turn, communicate those results to national decision makers and COP15 delegates' (DBT, 2008). DBT's media strategy document implies the same (DBT, 2009a, p1) and the WWViews project timeline features media dissemination prominently (DBT, 2009d, p1).

Discussions at the national level, at least in the US, mirrored DBT's emphasis on media. As researchers and volunteer staff for the Denver WWViews site, we were part of the many meetings and conference calls that took place around event planning in the US, and our notes from those meetings confirm that there was a general sense among planners (including us) that (1) media outreach was necessary and desirable for WWViews as a project and (2) generating media attention would be a fairly straight-forward means of attracting attention of policy-makers involved in the COP15 process.

In sum, WWViews organizers at the global and national levels put significant energy into media strategies and activities. While DBT had some noteworthy successes (see above), US coverage of WWViews did not appear to reflect the level of priority that national organizers gave to media strategy. We turn now to the task of explaining this discrepancy.

External media context: Climate as conflict in US media

Understanding climate change coverage in the US is complex because of the peculiarities of the US media and political systems and their relationship to climate change as an issue. It is well understood in the US that climate change – as a subject of public and political debate – is a challenging public policy issue for a number of reasons (e.g. see Moser and Dilling, 2007; Hulme, 2009; Leiserowitz et al, 2010). Polls and academic studies have revealed that public beliefs about the origins of climate change and, frequently, policy responses to climate change, frequently fall along partisan lines in the US (Dilling and Moser, 2007; Pew Research Center, 2009).

Furthermore, media coverage of the issue, until recently, has often appeared in the form of conflict-driven narratives that emphasized scientific disagreement over the causes of climate change, political wrangling over policy options, and – particularly in the blogosphere – sometimes vicious personal attacks among those of differing views.[1]

Although a thorough review of the politics of climate change media coverage in the US is beyond the scope of this chapter, it is worthwhile emphasizing that, for better or worse, WWViews was a climate change story (not a deliberative democracy story), which demands an understanding of the larger US media context in which climate change-related stories appear. Figure 14.1, produced by media scholars Boykoff and Mansfield, tracks the coverage of climate change in the 'prestige press' in multiple

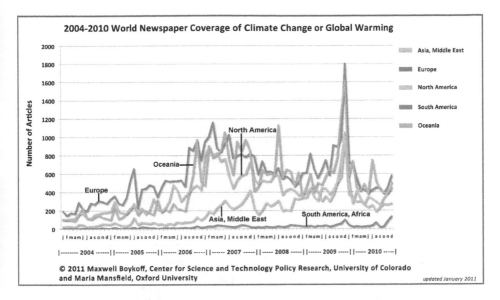

Figure 14.1 *World newspaper coverage of climate change or global warming 2004–2010*

Source: Boykoff and Mansfield (2011)

geographic locations around the world. The graph suggests a spike in global warming coverage around 2006 and 2007, which may be attributed to the popularity and attention given to Al Gore's film *An Inconvenient Truth* and the fact that he and the Intergovernmental Panel on Climate Change (IPCC) were awarded the Nobel Peace Prize shortly afterwards (see Robert Brulle, quoted in Revkin, 2010). Another significant spike occurred in December 2009, most likely due to the climate talks in Copenhagen and 'Climategate', a so-called scandal in which hacked emails from climate scientists revealed alleged wrongdoing or at least bias in the climate science community (see Boykoff as quoted in Yulsman, 2010).

If we narrow our focus to media coverage of climate in the US, an even more revealing picture of the media landscape emerges – one in which climate change coverage hardly features.

Figure 14.2 features news stories tracked by journalism.org, a group sponsored by the Pew Research Center. It includes stories in the US media extending through mid-December 2009, capturing stories related to 'Climategate', but not including a number of stories that followed the close of the COP15 meetings on 18 December. It also does not reflect the substantial and ongoing debates over climate change in the blogosphere (Revkin, 2010). Even so, according to journalism.org, climate change only received 1.5 per cent of all traditional news coverage in 2009 (quoted in Revkin, 2010). This number would have been lower if not for Climategate. For obvious reasons, the media

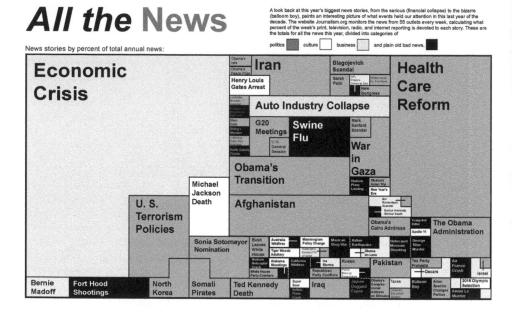

Figure 14.2 *Fifty-five outlets (internet, print, radio, television) monitored by journalism.org and compiled into an interactive map*

Note: Stories are represented graphically in proportion to the amount of coverage they received. Climate-related stories are nowhere to be found.
Source: Journalism.org, 2010. Original figure was made as a collaboration between GOOD, NAZ ŞAHIN and ŞERIFCAN ÖZCAN.

treat climate change as an 'environmental story', and the latter receive little or cyclical coverage when examined over time (see Brulle in Revkin, 2010; see also Anderson, 1997; Cox, 2006). Furthermore, as Revkin and many communication scholars have noted, we cannot rely on media to have particular impacts: media exposure cannot predict particular behaviours or attention, whether from the public or policy-makers (Anderson, 1997; Cox, 2006; Revkin 2010).

In sum, the US media presented a particularly challenging context for WWViews coverage, given its patterns of coverage of the environment and climate change, and the partisan character of climate change debates.

The 'Mediagenicity'[2] of WWViews in the US

It still makes sense that WWViews planners would have prioritized media coverage as an important 'impact' or 'dissemination' goal for the project. The project met some accepted criteria for being newsworthy in the US, and perhaps internationally: it was both large and continuous with other, ongoing stories. WWViews was novel because of its size – citizen deliberations had occurred before, but never on the scale of WWViews. The deliberation also met the criteria for continuity: climate change is an ongoing story, and COP15 an important policy event. According to Anderson, who has summarized many of the key maxims of environmental journalism, 'If an issue or event has already commanded media attention there is a greater likelihood that it will be continued to be viewed as newsworthy' (Anderson, 1997, p119). On this count, WWViews would seem to have passed the test for meriting coverage.

Unfortunately, the WWViews deliberation failed to meet several other criteria for newsworthiness. Environmental communication scholars have done extensive work on environmental 'image-events' (see e.g. DeLuca, 2006) and have detailed multiple news 'norms' that environmentally related stories must meet to break through. Most importantly, news events are narrative-driven, event-centred, meaningful to the audience in some way, and lack ambiguity. A number of other norms are discussed at length in Anderson (1997) and would be familiar to most public relations or communication professionals (Anderson, 1997, pp118–119).

One of the most important of these norms for climate change stories is conflict. Though not invariably, reporters and editors tend to frame the news in terms of conflict. Conflict guarantees emotional stories and ease of presenting both sides of an argument so as to showcase the neutrality historically fetishized by the media (Boykoff and Boykoff, 2004; Schneider, 2010). While addressing a policy conflict, WWViews as an event did not embody that conflict in its process; in fact, organizers imagined that it would be an event that would 'generate media coverage that contests stale "pro" and "con" narratives by presenting the results of deliberations by diverse groups of everyday citizens' (Worthington, 2008). Unfortunately, stale advantage and disadvantage narratives had thus far typified the climate debate in the US, and WWViews did not meet enough other news-making criteria discussed above to overcome this deeply entrenched, conflict-driven framing of the issue.

Breaking through: The Pew poll peg

Despite the challenges posed by the lack of 'mediagenicity' of WWViews, there was a media breakthrough of sorts in the US context: the appearance and reappearances of a blog post by WWViews organizer Richard Sclove. Sclove's (2009a) post, entitled 'Why the polls on climate change are wrong', initially appeared in late October and then was replicated or cross-posted numerous times on green blogs such as Treehugger (www. treehugger.com), the *New York Times* blog Green Inc. (http://green.blogs.nytimes.com), *Time* magazine online (www.time.com), Common Dreams (www.commondreams.org) and Grist (www.grist.org).

As the title of the post implies, the story pegged itself (meaning that it connected itself to an already ongoing story) to a particular narrative that had recently emerged with the release of a Pew Center poll on climate change.[3] Pew's results indicated that Americans' interest in or concern about climate change had fallen nearly 20 per cent in two years (Pew Research Center, 2009). The Pew poll received significant attention in the mass media and the climate blogosphere, seen by bloggers and others as evidence that climate change was indeed cause for little concern and the public knew it (e.g. Morano, 2009), or as evidence that we must redouble our efforts to educate the public about the severity of the crisis (e.g. Hoggan, 2009). Sclove's opinion piece, originally appearing in *The Huffington Post* (a progressive-leaning blog), pegged WWViews to this highly visible poll by arguing that WWViews, as an 'informed opinion poll', directly challenged the results of the Pew poll (Sclove, 2009a). Late October 2009 saw follow-up coverage that explicitly referenced Sclove's framing of WWViews as an 'informed opinion poll': an essay by Sclove that appeared in *Yes! Magazine* online (Sclove, 2009b), an independently written piece by the Worldwatch Institute (Block, 2009), extensive coverage of WWViews by Sea Change Radio (2009) and a blog post for PRI's syndicated show *The World* (Grossman, 2009). All are niche outlets that cater primarily to the environmentally minded, but all have some national and/or international exposure – a breakthrough into one level of media but still falling short of the kind of mainstream attention that many WWViews organizers had hoped for in the US. All these coverage examples follow the original Sclove framing of WWViews as an informed poll.

If we primarily value quantity of media coverage as a metric for success, the Pew poll peg approach succeeded. More than any other efforts by WWViews site organizers in the US, the 'informed opinion poll' comparison yielded significant attention, particularly in the blogosphere where climate change debates most frequently appear in the US. These debates are frequently intensely partisan and hyperbolic; therefore, breaking through so much 'climate noise' to feature citizen deliberation was not insignificant.

From a qualitative perspective, however, we wonder about the nature of the coverage garnered by the Pew poll peg approach and the risks of breaking through by emphasizing conflict. Again, we point to an early stated goal of WWViews organizers, which was that WWViews would somehow contest 'pro' and 'con' narratives about climate change (Worthington, 2008). By presenting WWViews as an 'informed poll' that stood in contrast to the Pew Center's 'uninformed poll', WWViews' organizers may have unwittingly recreated a stale 'pro and con narrative' in which US citizens were represented as fairly uniformly supporting one policy position, an assumption which is still being debated by communication and other scholars who have examined this issue closely.[4] In particular, the Pew poll peg limited the WWViews narrative to a single dimension that captured the difference between the 'informed' and 'uninformed' polls – the degree to which Americans were concerned about climate change – which of course failed to incorporate the suite of complex questions upon which WWViews participants had deliberated (e.g. the relative responsibilities of rich and poor nations for reducing greenhouse gas emissions, institutional requirements for technology transfer, and financing transitions to renewable energies). Furthermore, because the Sclove piece and others like it broke through on so-called green or liberal blogs, WWViews risked being seen as a partisan event, one in which the rooms of deliberators across the country and around the world were loaded with participants in favour of taking strong action on climate change, a typically liberal or progressive position in the US. It is important to note that the 'informed poll' peg was not created out of convenience or in response to a unique opportunity (the publication of the Pew poll). Even before WWViews took place, organizers imagined the event as akin to those organized and executed by James Fishkin, who has organized numerous, massive 'informed polls' on a range of issues around the world (Fishkin, 2009). Yet we would argue that framing WWViews as an informed poll was perilous when it came to the media context in the US. Although WWViews organizers emphasized the goal of achieving demographic representativeness within each deliberation, US organizers did not screen applicants for political affiliation, a key indicator of diversity for climate issues in the US. WWViews may have approximated an informed poll, but the majority of those polled on the day itself already cared about or understood climate change issues sufficiently to have chosen to be in the room in the first place (for an analysis of incentives to participate in citizen deliberations, see Kleinman and Powell, 2007).[5] In fact, the experiences of the US organizers suggest that this was a question frequently posed by reporters: 'Can you prove that the results from WWViews are representative?'

In an interview, DBT staff made the compelling argument that opinion polls have their own methodological shortcomings (rarely discussed publicly or explicitly) and that even national elections do not guarantee representative opinions of the population (some groups are more likely to vote than others). Such defences of WWViews did not gain traction in the US context for perhaps many reasons, and the Pew poll peg probably exacerbated anxiety over WWViews' ability to be representative, given the direct comparison to a methodology – the public opinion poll – generally accepted as reliably and objectively taking the pulse of the country on key political issues.

A different model for media coverage: Citizen outreach and institutional networking

Another model for thinking about success in terms of media has less to do with breaking through into national media dialogue on climate policy and more to do with citizen outreach and institutional networking. In the US, this model was embodied by the WWViews team that ran the event at the Museum of Science, Boston.

While Boston WWViews organizers explicitly sought media coverage (and had greater expertise in media relations than many of the university-sponsored sites in the US), they did so in the context of organizing several events beyond the standard day of deliberations. They were able to do so by drawing on their significant history of organizing such citizen- and community-oriented events, and by catalysing a network of citizens and experts already in place. First, they sponsored a forum the day after WWViews at the Museum of Science that included WWViews participants and experts who discussed the relationship between WWViews and the COP15 negotiations. This forum permitted reflection by WWViews participants on their experience, direct interaction with experts, and exposure of the WWViews project to museum-goers on that particular Sunday. Such a model – in which citizens are invited and empowered to participate in associated events beyond the deliberation itself – is one that future organizers of WWViews-type events may want to consider.

Second, as members of the international Association of Science and Technology Centers (ASTC) and of the Expert and Citizen Assessment of Science and Technology (ECAST) network, Boston WWViews organizers held a forum on 5 December that connected their local audience of 60 people to a similar event held at Cité des Sciences et de l'Industrie, Paris (a French museum of science and technology) with 400 attendees, and to WWViews lead organizers from DBT in Copenhagen (via webcast). All three sites attracted principal actors in climate change science and policy.[6]

Interviews with a Boston WWViews organizer demonstrate that the museum was careful in managing media attention, attaching some preference to sending information and releases to trusted sources only, given the severe partisanship that emerged with 'Climategate'. National Public Radio showed some interest but did not attend any of the events, but *Le Monde* (a French newspaper) and Sea Change Radio did send reporters to the 5 December event. Finally, through a funding organization that helped to sponsor the Boston WWViews event, organizers published several short articles about WWViews in a blog affiliated with a group of local newspapers (see e.g. Scammell, 2009).

In contrast to the disappointment felt by some US WWViews organizers regarding media coverage, the Boston WWViews coordinator for media and outreach expressed satisfaction for what the Boston organizers had accomplished. With greater attention to direct outreach to citizens and to networking with other significant institutions, not only did Boston organizers 'get the word out', but in recruiting key players to participate in forums, they also gained an audience of the very kinds of decision-makers that traditional media were supposed to access within the US WWViews media strategy.

We note, however, that replicating the Boston WWViews model would necessitate a very different framing of media and outreach at the programmatic level, and perhaps the recruitment of other science museums to play the role that universities have typically played in sponsoring deliberative forums in the US. We analyse this alternative model more closely in our conclusion.

Internal media context: Making the day happen

In addition to understanding the external media context in which WWViews was situated, it is also important to understand the obstacles and opportunities that event planners faced within the project. Generally speaking, US site organizers were a dedicated and competent group who organized these massive deliberations seamlessly and effectively. They were limited, however, in their abilities to seek and secure media coverage for a number of reasons.

Small staffs, small budgets, little time

From the beginning, one challenge faced by site and project organizers was a lack of funding. Initially, planners envisioned up to ten US sites participating in WWViews, but the inability to secure sufficient funds halved this number. Most US partner sites lacked the luxury to fund media outreach; instead, the priority for most sites was simply to get enough funding to make the deliberations themselves happen.

A second limitation was that most sites were organizing the WWViews event with small groups of academics and support staff who received little additional pay or formal recognition. For many academics, organizing any interaction with the public is treated as service, which is typically not valued as highly as research and teaching in the US university system. At some universities, WWViews sites had only one coordinator, with only graduate students or volunteer help. As one site coordinator put it:

> *Then the 26th [of September] came, and we were just totally focused on operation-alizing that day. And all of us were working on a limited amount of resources. When the economy is bad, all the democratic values get compromised first. So we had to work on a very limited amount of money, and still had to make it happen. We never had time [or a] moment to think about media.*

Some US sites had more significant resources to expend on media strategy, but all such efforts were secondary to making the event itself happen.

Furthermore, university public relations offices were frequently at a loss when it came to publicizing such an unusual event. As one public relations employee told us in an interview:

> *I thought this kind of event lent itself to a lot of social media, you could get some-thing going on Facebook or Twitter. But academics are driving the event, and you're on their timeline and comfort level ... If you're going to make it successful, you want to recruit people, and you want to have a concerted, coordinated marketing effort. You've got everybody working the angles of it with various constituents, outreach going out that touched people already interested in this issue, instead of starting from scratch.*

This comment reveals the difficulty media personnel faced in coordinating such a large effort and in trying to package the event in a way that would be palatable to the academics and others involved.

Other unforeseen conflicts arose as well. At one university, the administration ac-tively discouraged the event itself, fearing that it was a partisan effort to promote a particular policy position. They would not endorse the event, made it difficult to find a location for hosting it, and would not send any administrators to welcome the group of citizen deliberators. Another university, concerned about protecting the anonymity of the WWViews participants, would not allow cameras or other recording equipment in the room. Such constraints made seeking media coverage particularly difficult.

Lack of experience and training in media outreach

Reviewing the notes from the DBT-organized March 2009 training session and media plans from WWViews sites in the US and other countries located primarily in northern Europe, we found that media plans focused primarily on whom to contact. The US plans in particular were mostly lists of local, regional or national contacts – from local newspapers, to CNN, to particular prominent journalists – who might be interested in covering an event like WWViews (US National Planners, 2009). Indeed, WWViews organizers and event planners made valiant efforts to promote WWViews to these contacts. However, few media plans focused on how to get coverage. There was little discussion of how to frame the WWViews story and no mention of pegs or 'news holes' (opportunities in the media landscape that might increase the chances of getting coverage). Including the authors of this chapter, those writing media plans had little experience with public relations.

The guidance from DBT on media coverage may have inadvertently led to this emphasis on contacts rather than strategic approaches. In a January 2009 internal newsletter, DBT organizers encouraged national partners to first make a list of important media contacts, and second to identify 'your national media discourse' (DBT, 2009a). Almost all US and many international sites compiled some form of media plan, but most organizers in the US (including ourselves) focused primarily on the first request only. DBT's (2009a) own extensive *Media Strategy* document emphasizes messaging, framing, angles and possible audiences, but few local media plans did the same. Unfortunately, in the US there was no national media team to implement DBT's vision. A media consultant was eventually hired in October 2009, after the WWViews consultation, but before then most site organizers were primarily focused on planning for the event itself. To the best of our knowledge, there was little discussion about the national media discourse in the US and how best to engage it. This seemingly changed once the media consultant was hired: the result was the Pew poll peg and an upturn in coverage.

Also, US organizers were not agreed about what kind of coverage they should seek. During one conference call in August 2009, there was concern that pairing with environmental organizations such as the Sierra Club to ensure more media coverage or dissemination might colour the event with a partisan lens. Similarly, in the *US Media and Dissemination Plan*, organizers wrote, 'Joining with an established organization can lend WWViews added legitimacy and existing networks. We must be careful, however, to pursue potential partners keeping in mind WWViews' intended neutrality' (US National Planners, 2009). In fact, these concerns may have turned out to be prescient, as our discussion of the Pew poll peg above suggests. In any case, most US national partners experienced substantial frustration when the WWViews story was not picked up, despite their good efforts.

Moving forward: Suggestions for media planning in future WWViews

While the scope of this chapter is primarily limited to understanding media coverage of WWViews in the US context, we think it is important to acknowledge that media realities in other countries (like political realities) could have been quite distinct from our own, especially as they related to climate change. Furthermore, we acknowledge the possibility that a number of policy pathways for disseminating WWViews results may have been more or less available in other countries than they were to us in the US.

Though we have limited ability to make claims about particular media successes around the world, it might be helpful to briefly draw a global sketch of these as best we can in order to better contextualize our focus on the US: based on a number of informal reports, generally speaking, there was no large-scale, substantive media coverage of WWViews in the various host countries or elsewhere around the world. This is not to say there were not notable instances of media stories about the deliberation, nor do we mean to imply that significant efforts to garner media coverage were not made. For example, the coverage WWViews received in Denmark as a result of DBT's efforts (discussed above) was notable; also, the Mayor of Sydney, Australia, disseminated a press release on the event, which could be seen as politically significant (Herriman et al, Chapter 13, this volume; Moore, 2009). However, we are arguing that – given the limited coverage climate-related stories receive in the US and worldwide – it would have been very challenging for WWViews to break through and receive considerable coverage. Informal reports from DBT and self-reports from countries around the world suggest that WWViews received similar coverage to the US in terms of quantity and quality in northern European countries and Canada, with a few (albeit notable) hits in South America, Australia and Asia.

In retrospect, a few lessons from our experiences with media in the US emerge. The limited media coverage achieved by nearly all WWViews partners should give us pause, and encourage us to think about how we might move forward.

Identify and secure a 'media partner'

Given the success of one WWViews US site (Boston, discussed above) in partnering with a group of institutions that focused on different aspects of WWViews, the potential for including a media partner in WWViews is attractive. Such a partner might range from the modest resources of a science television programme (e.g. *Nova*) to an international news network (e.g. the BBC). As noted by DBT staff in interviews, the first WWViews now serves as 'proof of concept' and might enable a major media partner to engage with future processes modelled on WWViews.

Hire a media consultant

We realize, again, that it is not always possible to divert funds from event planning and coordination to media coordination, but in the US context the hiring of a media consultant marked a turning point in the type and amount of coverage. Media consultants can do much of the legwork academics or other volunteers cannot, and they frequently have connections with public relations professionals, reporters and other media outlets already in place. Perhaps most importantly, they can assist in strategizing appropriate news pegs or story framings.

Create an 'event in a box'

One of the public relations professionals we spoke to at a US university site suggested that planners come up with a sort of 'event in a box', a mediagenic, sponsored happening that could draw attention to the deliberations, because the deliberations themselves were not mediagenic. In an interview with us, she said, with an 'event in a box':

> ... you're giving someone a visual. Whereas for us, I just threw up my hands at one point, because what I have now are a bunch of people around tables in a room. The press can cover the mayor to make the opening statement, but ... there's no visual hook. Even when you're dealing with print press ... there has to be a physical event, a kick-off, and not just handing a report to somebody ... So the question becomes, how do you create something? [our emphasis]

While we would caution against a complete redesign of the WWViews deliberations to provide a compelling media hook, we imagine marginal changes to the format or the addition of other activities that would resonate with the needs of mainstream media.

Maintain flexibility across national and cultural contexts

At the same time that planners might consider 'boxed' events that can garner more media coverage than deliberations, they must always also be mindful of local contexts. If our chapter has done anything, we hope that it has exposed the particular media realities that surround climate change in the US. Similarly contextualized realities are no doubt important for all worldwide sites, though for different reasons.

Conclusion: Why media?

Our dual roles as WWViews volunteers on the one hand and as scholars studying the event on the other lead us in this chapter to both identify ways WWViews might better 'do' media coverage in future and, at the same time, question whether 'doing' media should be a priority.

One significant reason why we have been interested in these questions has to do with how citizen deliberation events recruit participants and instil the events with purpose for these participants. We have been involved in two citizen deliberation events now (including WWViews) and have noticed that in both projects similar promises were made to participants about the purpose their involvement would serve. In both cases, participants were told that their work on the project would affect policy-making in some way (for a discussion of such issues surrounding the National Citizens Technology Forum, see Kleinman and Powell, 2007). WWViews participants were frequently imagined as 'citizen advisers' (see e.g. Sclove, 2009b). For example, in California's 'invitation to participate', which served as a template for most US sites' recruitment strategies, potential participants were invited to:

> *Join participants from over 40 nations in a discussion on global warming and share your ideas with the United Nations! Your voice will be added to recommendations being developed for the 2009 United Nations Climate Change Conference in Copenhagen in December ... These global discussions offer participants the opportunity to be a part of an international conversation about environmental policy options that affect us both locally and globally.* (Pomana College, 2009)[7]

This letter was drafted some months before the event was held, and the amount of dissemination and policy impact was still incredibly uncertain. Yet there were fairly certain promises made here: that participants would be part of a global conversation (which ended up being true, to some extent) and that their voice would in some way be represented at COP15 (which came true in only a very limited sense). If one of the major points of a WWViews exercise is to 'have one's voice heard', then it makes sense that media outreach would seem valuable. If, however, the significant goal of citizen deliberation is the deliberation itself – the entering into a global conversation – then media outreach becomes less important.

This understanding of WWViews participants as 'citizen advisers' no doubt has roots in DBT's classic 'consensus conference' model (Joss, 1998; Sclove, 2000; Kleinman and Powell, 2007). In that model, citizens are invited by the Danish Parliament to learn about and debate science and technology policy. The citizens' recommendations on particular policy options are then considered and sometimes incorporated into Danish law-making. In this context, therefore, small groups of citizens do actually function as advisers to policy-makers. But the question of whether that is an appropriate or accurate model for WWViews is an open one. What we saw in the process of making WWViews deliberations happen in the US, in fact, was a tension between wanting citizens to function as 'citizen advisers', as they do in consensus conferences organized by DBT, and having them simply participate in the event to 'make their voices heard', which is a much more amorphous concept, and one not necessarily connected to particular policy pathways.

We would argue, therefore, that the two stated purposes of WWViews – citizen deliberation on the one hand and policy intervention on the other – may exist at cross-purposes, particularly when it comes to understanding the role of media coverage in the US context. National partners in the US were primarily concerned with making the WWViews consultation itself happen smoothly and with substantive deliberation. However, they did frame the event for participants within the larger policy context of making citizen voices heard at COP15. Unfortunately, as the weeks after the WWViews event unfolded, it became clear to the US organizers that having a policy impact in Copenhagen would be neither easy nor straightforward. US delegates to COP15 were first difficult to identify, and then very difficult to contact and interact with. Also, there was no guarantee that they would have taken an interest in WWViews if they were contacted. Finally, in hindsight, it is unclear how they could have used this information to impact the discussions at COP15, given that negotiating positions for various countries were determined well in advance.

Without a clear policy pathway for disseminating results to delegates at COP15, therefore, the pressure on media outreach in the US seemed to increase in the months following the WWViews deliberation. More efforts were made to reach local and national media outlets; but the window of opportunity for gathering media attention was small, and the internal infrastructure needed to build media interest had not been created.

In an October press release, WWViews organizers wrote:

> *The main objective of World Wide Views is to give a broad sample of citizens from across the earth the opportunity to influence global climate policy. An overarching purpose is to set an historic precedent, demonstrating that political decision making at the global level benefits when everyday people participate.* (DBT, 2009c)

We want to make it clear that, as participants in WWViews on global warming, we support this objective and overarching purpose. When it comes to problems of a particularly global nature, such as climate change, inviting citizen involvement in this way is admirable and worthy. At the same time, WWViews organizers may reflect further on the paradoxes of seeking media attention as a policy pathway. Garnering media attention frequently requires a conflict narrative, and emphasizing conflict may itself undermine the deliberative approach of the WWViews model. Instead, focusing on more direct policy pathways – such as partnerships with policy-makers or institutions, or emphasizing public outreach and education, rather than public attention, per se – might serve to complement the deliberative values embodied in a WWViews process.

Acknowledgements

This research was supported by a grant from the US National Science Foundation (World Wide Views on Global Warming: Process and Outcomes, NSF Award #0925043). Awardees included researchers from Pomona College, Arizona State University, Colorado School of Mines and Georgia Institute of Technology. In addition, we acknowledge the key role that other US partners played in carrying out additional WWViews deliberations and contributing to our understanding of the process: Museum of Science-Boston, Boston University, The Brookfield Institute and The Loka Institute.

Notes

1   There are a number of climate-related blogs that provide evidence of this; perhaps most telling is the ongoing online debate between Joseph Romm of the blog Climate Progress (www. climateprogress.org) and Roger Pielke Jr, whose blog can be found at www.rogerpielkejr. blogspot.com.

2   WWViews planners and their media consultants in the US often used the colloquial term 'mediagenic' to refer to an event that carried a strong media appeal in its basic design, or genes, and frequently noted that WWViews lacked this quality.

3   A peg is a short rod (normally made of wood) in a wall from which clothes or a hat can be hung. The idea of 'pegging' a story is commonly used among US media specialists to describe the process of connecting a specific story to a larger frame of reference in the news so that the story will be meaningful to readers, listeners and viewers.

4   For a wide-ranging study of the range of public attitudes towards climate change in the US, see the reports from George Mason University's Center for Climate Change Communication – in particular, their recent report *Climate Change in the American Mind* (Leiserowitz et al, 2010).

5   There are certainly exceptions to this claim. After the event, we interviewed a number of climate sceptics who had participated in WWViews. Though we cannot generalize to all US sites, the sceptics we spoke to did not feel that the event was open to debate concerning their views on the science of climate change. Indeed, WWViews project organizers made clear that the scientific uncertainty around climate change was not a focus of WWViews; instead, WWViews took for granted the same assumptions as would underlie the COP15 negotiations – specifically the scientific consensus presented by IPCC – in order to focus on deliberations over policy responses.

6   The impressive group of participants included: Jean-Pascal van Ypersele (deputy chair of IPCC) and Trinto Mugangu (COP15 delegate from the Democratic Republic of Congo) in Copenhagen; Sandrine Mathy (president of Climate Action Network) and Stéphane Hallegatte and Jean-Charles Hourcade (representatives from the International Research Center on Environment and Development) in Paris; and Peter Schultz (former director of the US Climate Change Science Program Office), Anthony 'Bud' Rock (ASTC CEO and former US principal deputy assistant secretary of state for oceans, environment, and science), Henry 'Jake' Jacoby (co-director of the MIT Joint Program on the Science and Policy of Global Change), and Bob Corell (chair of the Climate Action Initiative and global change program director at the H. Heinz Center for Science, Economics, and the Environment) in Boston. Readers who are interested in more information on ECAST can go to www.ecastnetwork.org. Of particular interest is Richard Sclove's report *Reinventing Technology Assessment: A 21st Century Model*, which can be found at http://wilsoncenter.org/topics/docs/ReinventingTechnologyAssessment1.pdf.

7   Such language was not confined to recruiting materials and was representative of how the event was publicized. For example, a press release sent from our own university site (Denver) claimed, 'One hundred Colorado citizens will join over 4000 citizens from around the world and will give political leaders some hints about what ordinary citizens think of climate change. Coloradoans will make their voices heard at a citizen meeting ...' (Colorado School of Mines, 2009).

References

Anderson, A. (1997) *Media, Culture and the Environment*, UCL Press, London
Block, B. (2009) 'US Public Still Unconvinced on Climate Change', The WorldWatch Institute, 26 October, www.worldwatch.org/node/6300, accessed 22 February 2010
Boykoff, J. and Boykoff, M. (2004) 'Journalistic bias as global warming bias: Creating controversy where science finds consensus', *Extra! The Magazine of Fairness and Accuracy in Reporting*, www.fair.org/index.php?page=1978, accessed 23 April, 2011
Boykoff, M. and Mansfield, M. (2011) '2004–2010 world newspaper coverage of climate change or global warming', University of Colorado at Boulder, Center for Science and Technology Policy Research, http://sciencepolicy.colorado.edu/media_coverage/
Burawoy, M., Burton, A., Ferguson, A. A., Fox, K., Gamson, J., Gartrell, N., Hurst, L., Kurzman, C., Salzinger, L., Schiffman, J. and Ui, S. (1991) *Ethnography Unbound: Power and Resistance in the Modern Metropolis*, University of California Press, Berkeley
Colorado School of Mines (2009) *Citizens Advise Politicians on Climate Change*, Press Release, Colorado School of Mines, Golden, CO
Cox, R. (2006) *Environmental Communication and the Public Sphere*, Sage, Thousand Oaks, CA
DBT (Danish Board of Technology) (2008) *Strategy for Achieving Policy Impact, World Wide Views on Global Warming*, Danish Board of Technology, Copenhagen
DBT (2009a) *Media Strategy for World Wide Views on Global Warming*, Danish Board of Technology, Copenhagen
DBT (2009b) *Policy Report, World Wide Views on Global Warming: From the World's Citizens to the Climate Policy-Makers*, Danish Board of Technology, Copenhagen
DBT (2009c) *Press Release*, Danish Board of Technology, Copenhagen
DBT (2009d) *Timeline with Tasks for National WWViews Partners*, Danish Board of Technology, Copenhagen
DeLuca, K. M. (2006) *Image Politics: The New Rhetoric of Environmental Activism*, The Guilford Press, Mahwah, NJ
Denzin, N. K. and Lincoln, Y. S. (2008) *Strategies of Qualitative Inquiry*, 3rd edn, Sage, Thousand Oaks, CA
Dilling, L. and Moser, S. C. (2007) 'Introduction', in S. C. Moser and L. Dilling (eds) *Creating a Climate for Change: Communicating Climate Change and Facilitating Social Change*, Cambridge University Press, Cambridge
Fishkin, J. (2009) 'Town halls by invitation', *The New York Times*, 15 August, www.nytimes.com/2009/08/16/opinion/16fishkin.html, accessed 22 February 2010
Frey, L. R., Botan, C. H. and Kreps, G. L. (2000) *Investigating Communication: An Introduction to Research Methods*, Allyn and Bacon, Boston, MA
Grossman, D. (2009), 'Parsing global climate change polls', *PRI's The World*, 14 December, www.theworld.org/2009/12/parsing-global-climate-change-polls, accessed 23 April, 2011.
Hoggan, J. (2009) 'New Pew Center poll confirms the effects of climate confusion campaign', *DeSmogBlog.com*, 22 October, www.desmogblog.com/new-pew-center-poll-confirms-effects-climate-confusion-campaign, accessed 22 February 2010
Hulme, M. (2009) *Why We Disagree about Climate Change: Understanding, Controversy, Inaction and Opportunity*, Cambridge University Press, Cambridge

Joss, S. (1998) 'Danish consensus conferences as a model of participatory technology assessment: An impact study of consensus conferences on Danish Parliament and Danish public debate', *Science and Public Policy*, vol 25, no 1, pp2–22

Journalism.org (2010) *All the News: News Stories by Percent of Total Annual News, 2009*, Journalism.org. Transparency posted by *Good Magazine,* http://awesome.good.is/transparency/web/0912/all-the-news/flat.html, accessed 7 February 2011

Kleinman, D. L. and Powell, M. (2007) 'A toolkit for democratizing science and technology policy: The practical mechanics of organizing a consensus conference', *Bulletin of Science, Technology & Society,* vol 27, no 2, pp154–169

Leiserowitz, A., Maibach, E. and Roser-Renouf, C. (2010) *Climate Change in the American Mind: America's Global Warming Beliefs and Attitudes in January 2010*, Yale University and George Mason University, New Haven, CT

Lindlof, T. R. and Taylor, B. C. (2002) *Qualitative Communication Research Methods*, Sage, Thousand Oaks, CA

Moore, C. (2009) *Parliament Out of Step with Public Opinion on Climate Change,* Press Release, Lord Mayor of Sydney, 23 November, www.wwviews.org/files/WORLD%20VIEWS%20POLLING%20ON%20CLIMATE%20CHANGE%20MEDIA%20RELEASE.pdf, accessed 29 December 2010

Morano, M. (2009) 'Updated: Climate depot factsheet on public opinion about global warming: Americans growing increasingly skeptical', *Climate Depot Blog*, 22 September, www.climatedepot.com/a/3025/Climate-Depot-Factsheet-on-Public-Opinion-About-Global-Warming-Americans-Growing-Increasingly-Skeptical, accessed 22 February 2010

Moser, S. C. and Dilling, L. (eds) (2007) *Creating a Climate for Change: Communicating Climate Change and Facilitating Social Change*, Cambridge University Press, Cambridge, UK

Pew Research Center (2009) *Fewer Americans See Solid Evidence of Global Warming: Modest Support for 'Cap and Trade' Policy*, Survey Reports, The Pew Research Center for People and the Press, Washington, DC, http://people-press.org/report/556/global-warming, accessed 22 October 2009

Pomana College (2009) *World Wide Views Invitation to Participants*, Pomona College, Claremont, CA

Revkin, A. (2010) 'The greatest story rarely told', *The New York Times Online*, 2 January, http://dotearth.blogs.nytimes.com/2010/01/02/the-greatest-story-rarely-told, accessed 22 February 2010

Scammell, M. L. (2009)'World Wide Views on Global Warming: Highlighting the results', *The Public Humanist Blog*, 22 December, www.valleyadvocate.com/blogs/home.cfm?aid=11055, accessed 25 February 2010

Schneider, J. (2010) 'Making space for the "Nuances of Truth": Communication and uncertainty at an environmental journalists' workshop, *Science Communication*, vol 32, no 2, pp171–201

Sclove, R. E. (2000) 'Town meetings on technology: Consensus conferences as democratic participation', in D. L. Kleinman (ed) *Science, Technology, and Democracy*, State University of New York Press, Albany, NY

Sclove, R. (2009a) 'Why the polls on climate change are wrong', *The Huffington Post Blog,* 23 October, www.huffingtonpost.com/richard-sclove-phd/why-the-polls-on-climate_b_331896.html, accessed 22 February 2010

Sclove, R. (2009b) 'World's Citizens to Politicians: Get Serious on Global Warming Now!' *Yes! Magazine Online*, 29 October

Sea Change Radio (2009) 'How everyday folks world wide view climate change', 30 September, www.cchange.net/2009/09/30/how-everyday-folks-world-wide-view-climate-change, accessed 22 February 2010

US National Partners (2009) *US Media and Dissemination Plan, World Wide Views on Global Warming*, US National Planners

Worthington, R. (2008) *World Wide Views on Global Warming (US Team): Organizing US Participation in a Global Citizen Consultation on Climate Policy*, Claremont, CA

Yulsman, T. (2010) 'Striking spike in global warming coverage', CEJournal, 16 January, www.cejournal.net/?p=2726, accessed 22 February 2010

Part VI
Conclusions

15

Deliberative Global Governance: Next Steps in an Emerging Practice

Richard Worthington, Mikko Rask and Birgit Jæger

In this concluding chapter we will take a step back from the diverse and detailed engagements with WWViews presented in the principal contributions to this volume in order to identify patterns and implications that emerge from the body of work as a whole. The chapters to this point mostly examine WWViews in the national contexts where the deliberations were conducted, or assume a national perspective when discussing international aspects of the project. However, recalling the desire to address global problems that originally motivated the Danish Board of Technology (DBT) to launch the project, we will shift our focus in this chapter to the lessons learned and possibilities created at a global level.

First we will discuss what deliberative global governance is. How can we understand this term in a theoretical as well as a practical way? After this we will examine the dynamics of developing global deliberations; the stakes for key players in such deliberations and how these affected the success and adaptation of WWViews in diverse settings; ideas for the future development of global deliberations; and important areas for continued research.

Characteristics and roles of deliberative global governance

Global governance is often equated with international institutions like the United Nations and the World Bank that were established after the Second World War to solve problems that transcend national borders. For the most part, these institutions project to an international level the norms of representative democracy that evolved in national societies: nation states thus play in international institutions the role of representatives that are occupied by individuals (elected and otherwise) in the national polity. This establishes a key distinction in governance at these two levels, which is that global institutions are comprised of sovereign representatives, while the national state is sovereign. As a consequence, global institutions cannot exercise over the issues that come before them the sovereignty that their members enjoy with respect to national issues. Global governance therefore confronts issues that are more complicated because of their global scope, while its capacity to address these problems is limited by what Raymond Aron has called 'the anarchical order of power' at the world level (Aron, 1995). As we know now, COP15 was a significant example of this complex and compromised international decision-making process that is derived from and limited by competing sovereign states.

Bexell et al (2010) report that international institutions are seen as lacking legitimacy due to the limited problem-solving effectiveness that stems from the challenges just reviewed, as well as a failure to establish more democratic procedures of decision-making (p11). Presumably the first criticism requires no explanation in relation to COP15, but the second calls for further elaboration. Here the focus is on the input to decision-making and the process itself, enhancing the prospects that deliberative democracy will increasingly be seen as a complement to strictly representative models. As described in several chapters, theories of deliberative democracy argue that deliberation among ordinary citizens, and between them and politicians, can confer legitimacy on the decisions that take such deliberations into account (Dryzek, 2006).

To ensure the success of deliberation, two preconditions have to be fulfilled: access to reliable information is important, as is the existence of a public sphere (Bexell et al, 2010). These conditions are even more difficult to fulfil at a global level than at a national level. In the case of climate change, the sheer complexity of the issue makes access to reliable information a significant challenge.

The second precondition of a public sphere is also difficult to fulfil at a global level for the reasons we have already noted. And for ordinary citizens it is no exaggeration to say that a 'global public sphere' does not exist. Even though electronic media like television, the internet and social media now reach beyond national boundaries, and are increasingly sites where global issues are discussed, in themselves they do not form a 'global public' space where ordinary citizens can exchange information, where deliberation is an accepted feature of public discourse, where opinions are shaped, and finally where decisions are made. This leaves global problems to be addressed by the anarchical order of power derived from competing states, or alternatively, given the limited effectiveness of that 'order', in the private sphere dominated by global (media) corporations. As a result, the most important concern of global governance, humanity's common fate, has no venue for meaningful consideration and informed action.

Within this framework, however, Bexell et al (2010, p2) show that transnational actors such as non-governmental organizations (NGOs), advocacy networks, party associations and transnational corporations have increased their participation in global governance. The authors focus primarily on the modest record and difficult trade-offs experienced by advocates for democratization of global governance, but they also show that ordinary citizens are able to influence decision-making through their participation in deliberations if they organize and act as a transnational actor.

In this perspective it is possible to understand the whole WWViews exercise as DBT's way to organize ordinary citizens and shape a kind of transnational actor whose voice is heard in both national deliberations and international decision-making bodies such as that United Nations Framework Convention on Climate Change (UNFCCC). Thus the results of WWViews have the potential to act like 'the voice of the ordinary citizens' and play a role as a transnational actor in the deliberation of climate change. Here we do not conceptualize the results solely as the final report from WWViews, but also as the network of active partners (the WWViews Alliance) who arranged the event in various local settings. In the long run, this network is probably a more important result of WWViews because it has the potential to act as an organized transnational actor able to bring 'the voice of the ordinary citizens' into global deliberative governance in relation to a broad range of issues in addition to climate change.

A key conclusion by Bexell et al (2010) is that transnational actors increasingly play a role in the process of democratizing global governance. They write: '… we find considerable support for an optimistic verdict on the democratizing potential of greater transnational actor involvement. Most notably, bringing on board NGOs, social movements, and advocacy networks can expand participation in global governance …' (p4). Despite this optimistic conclusion they also point at some pitfalls. Transnational activism often lacks participation from the global South, just as structural inequalities based on class, gender, ethnicity and religion are often reproduced in the more affluent countries where deliberative democracy has grown in recent decades (p27).

As we saw in the chapter by Bedsted, et al., DBT designed a process for recruiting diverse participants into all its citizen deliberations in order to construct a legitimate voice. DBT was also aware of the structural inequalities that deter participation from the global South in international processes, and actively cultivated alliances there while seeking funding to support those who joined. Despite the global financial crisis that peaked during this period, funds were secured from international sources to support partners in the global South, who ultimately constituted 31 of the 65 members in the WWViews Alliance.[1] Much like the challenges of representation in deliberative events discussed in several chapters, the global South was under-represented in the world's first global citizen consultation, largely because of greater resource constraints. In instances of structural inequalities, however – both those within national deliberations and those among the deliberating nations represented at a global level – WWViews can be seen as engaging the democratic challenges that are a key to the legitimacy of global governance with a level of success that would compare quite favourably with other transnational phenomena.

In conclusion, 'the voice of the ordinary citizens' has the potential for playing a part in the process of democratizing global governance. Until now this voice has been weak due to the ephemeral nature of the 'global public sphere'. The ways citizens can be heard in global governance for the time being are by electing national politicians who take part in international negotiations or by organizing as transnational actors such as social movements, NGOs and the like. But for citizens who lack the desire or opportunity to organize as transnational actors, it is in practice impossible to gain a voice in global governance. From this perspective the work of DBT and the WWViews Alliance is uniquely important. By arranging WWViews, these players helped disorganized ordinary citizens achieve a voice in global deliberative governance. And more important: the WWViews Alliance has the potential to become a transnational actor with the aim of institutionalizing global deliberation.

The dynamics of developing global deliberations

How is the space for global citizen deliberation evolving in practice? This question can be approached from the perspectives of methodological evolution (micro dynamics), organizational learning and interaction (meso dynamics), and socio-economic and cultural contexts (macro dynamics). All three perspectives reveal significant facets of an emerging field of practice, but building on our observation above about the potential role of the WWViews Alliance, we will argue in this section that the most significant opportunities to shape future global deliberations are found at the meso level.

Methodological evolution

The evolution of global deliberation is nicely depicted by Bedsted et al[2] in their account of DBT's growing appreciation that many urgent problems transcend the national scope of most deliberations to date, and in Andersson and Shahrokh's account of the progression of deliberative events from national to European to global levels in recent years.[3] Addressing the micro level of this evolution, DBT describes the method created for WWViews as a hybrid based on its own experience and those of WWViews Alliance members and others over a period of several decades (DBT, 2010a). Because both the method and the constellation of players putting it into practice are new, the implementation was marked by numerous judgement calls. While study and reflection can certainly strengthen the basis for such judgements in future, our sense is that this is likely to be an enduring feature of global deliberative methods.

The most vexing methodological issue was standardization of the voting questions. This drew little attention in relation to the citizen recommendations component of WWViews, but the standardization of voting to achieve comparable quantitative results was the subject of intense discussion on the project intranet, and continues to be debated in this volume and elsewhere.[4] In the process of developing WWViews, the standardization goal also pitted the desirability of common practices for recruitment, format, information provided to participants, etc., against adaptation to local circumstances. Some requests for slight variations from the standard practice were relatively easy to accommodate, such as a concern in Chile that a day-long process beginning at 9.00 am on a weekend was far too early in that nation's culture, while others, such as a US request to change the wording on one of the questions, were rejected by DBT in the interests of maintaining comparability of the deliberations at all sites.

One advantage of a gradual evolution of deliberative practices that builds on an increasingly rich body of experience and research is the wisdom it provides for constructing a methodologically informed and politically robust deliberation process. A gradual evolution would also provide a space where diverse and new participants can be on the same footing regarding the method. The development of deliberative global governance itself can therefore incorporate practices and norms of pTA, including face-to-face interaction among organizers and participants at multiple levels of planning, presenting results and reflecting on outcomes. In this light, standardization may be most significant not as a method for comparable results, but as the basis of a coherent conversation that can help establish a global public sphere. A deliberative process that is focused on the same questions and directed towards the same decision-making process is surely important in building on the pluralistic discourses of the nation-state system.

Organizational learning and interaction

The meso dynamics of organizational learning and interaction constitute a second level at which the development of global deliberations can be understood. Learning inside the WWViews Alliance was a crucial element of the project, especially in light of the fact that many partners had no prior experience with technology assessment or deliberative democracy. Even those more experienced in these fields were confronted with the challenges of understanding the new 'hybrid' method, establishing relationships with others in the network, learning the internal structure and processes of the project and developing an adequate working knowledge of international climate change policy.

In addressing these and other learning needs, DBT adopted the posture of the 'network extender' who builds value through openness and empowerment, in contrast to the more opportunistic 'networker' who deploys secrecy and cliques to concentrate profit while spreading costs and risks to others (Boltanski and Chiapello, 2005). This approach is evident in all aspects of DBT's leadership. For example, a central design criterion called for a 'cheap and easy' project that made it 'feasible for potentially all countries in the world to participate, regardless of financial income and general education level' (DBT, 2010b). DBT's efforts to secure funding for developing countries, the organization of a project manager training seminar and an intranet for ongoing collaboration and information sharing, and investment in a web tool to quickly report the deliberative results in a searchable format also speak to the centrality of transparency and empowerment in strengthening the WWViews Alliance for future deliberative events. An ongoing challenge will be remaining open to new participants and practices while adding sufficient stability to the network to reap the benefits of these expensive investments. Success in this regard will be crucial in ensuring policy impact in national and international bodies.

A second important arena of organizational learning and interaction occurred between the WWViews Alliance and national/international climate policy actors. Notwithstanding the global scope of climate change and UN negotiations to address it, the organizational reality of WWViews is clearly multilevel. The very design of the project calls for national deliberations that produce policy input for national leaders, who in turn participate in an international policy process. Even so, the WWViews Alliance itself participated directly at an international level, for example, by organizing a discussion of ambassadors from China, India, Japan, Uganda and the European Union in the Danish Parliament when its policy report was released, and by holding side events at the COP15 site and at Klimaforum, 'the people's' climate forum in Copenhagen. The German proposal (Goldschmidt et al) to recognize this multilevel quality by holding international deliberations among citizens selected from national events speaks to the need to directly address the complex layering in global deliberations. More suggestions, however, moved in the opposite direction by recommending the inclusion of information on local climate change issues in the deliberations in order to help citizens connect the global issues to their everyday realities.

The stakes involved in designing and implementing global deliberations that are in actuality events focused at multiple levels are significant. The state system is organized from sub-national through to international levels, but (as noted) is incapable of acting effectively on global problems. Transnational actors are unlikely to significantly change this situation merely by intervening at the top. Evolving a network that supports an authentic global citizen voice will therefore require continuous and significant organizational learning by any transnational actor who embraces this challenge.

Socio-economic and cultural contexts

Finally, we can view global deliberations at a macro level that examines the cultural and socio-economic contexts in which they develop. The most prominent socio-economic factor for WWViews was the combination of a severe global recession with the special challenges of a start-up venture. Entrepreneurial efforts rely far more on forging new connections and sources of support than they do on tapping existing relationships, and by definition they have to establish the value of a new product or service. Doing so during the worst economic recession since the Great Depression presented unique difficulties. Unless there is sustained global stagnation, the economic environment of future WWViews can be expected to be more favourable, and both the method and the WWViews Alliance will be better established.

Connecting discrete events and outcomes of complex processes is challenging in the best of circumstances, and arguably as difficult in the global climate change arena as for any other issue. Nonetheless, the Technology Assessment Methods and Impacts (TAMI) project conducted by European practitioners provides a framework that brings some order to this daunting but important task (Klüver et al, 2004). In this framework, the impacts of technology assessment (whether deliberative or not) can occur in the dimensions of raising knowledge, forming attitudes and opinions, and initiating policy actions.

Notwithstanding the small scale and novel quality of WWViews in the COP15 decision-making process, potential and actual outcomes can be identified in all three of these areas. While WWViews had a small footprint compared to media-focused events such as 350.org's day of action on 24 October 2009, it nonetheless directly involved more participants than prior international deliberations by multiple orders of magnitude (Sclove, 2010, p36). For the future, the important domain for improving awareness is helping to construct citizen voices that are distinctive in comparison to the arguments emanating from the advocacy and research organizations that currently prevail in civil society discourses, and that are focused on determining the 'right' policies to 'fix' the problem. Despite the value of these discourses, their legitimacy is delimited by the self-appointed nature of the groups engaged in them.

As to attitude formation, the votes and especially the recommendations from WWViews represent an impressive body of suggestions that can help set the agenda in political debate. Policy-makers interviewed several months after COP15 have encouraged the organizers to continue bringing these to the attention of national leaders.[5] As already argued, the most important action initiated to date is the formation of the WWViews network itself. Locally, however, national partners are following up in numerous ways, for example, through youth deliberations on climate and environmental issues envisioned in Saint Lucia (Charles et al).

Understanding the stakes in global deliberation

Many chapters in this volume describe the reactions of NGOs and campaigns to the WWViews process. Notwithstanding their overlapping agendas and interests, WWViews often competed with these organizations for resources and attention, while few of them used WWViews results as a resource in their advocacy. The high political stakes in international climate policy make the piloting of a global citizen deliberation process, especially one that aims to influence UN negotiations, an audacious and challenging move. However, other types of stakes – financial, scientific and reputational – are also important in the development of global deliberations.

It is important to recognize that the main stakeholder in the WWViews process is virtual: the global citizen. This citizenship is virtual, or at least tentative, because identity formation historically has emerged from common values, religions and experiences shared by communities of interest or members of states (Blue et al; Young, 1989), as well as from external forces such as the threats of neighbouring communities, or natural disasters that imperil daily subsistence. While the rise of modern threats such as nuclear proliferation, reduced biodiversity or global warming have expanded the geographic scope of environmental hazards, global citizenship remains unarticulated. This partly reflects the simple fact that for most people, everyday concerns remain locally based, both institutionally and perceptually, while transboundary issues seem remote despite their transmission into homes worldwide by television and the internet.

WWViews was partly motivated by DBT's conviction that these constraints could be loosened by providing citizens the opportunity to influence global climate policy, thus increasing their ownership of decisions at this level (Bedsted et al). As an internationally renowned practitioner of pTA, however, DBT had significant stakes in the development of WWViews in addition to these political outcomes. Their mobilization of international funding, political legitimacy and methodological skills to ensure the high quality and political impact of the project shows that DBT deployed its financial, reputational and scientific stakes in the mode of a network extender. In an environment where core processes in citizen participation have been privatized by some organizations (e.g. Citizens Jury®, Deliberative Polling® and 21st Century Town Meeting™),[6] this building of open and empowered networks, incorporating many partners who were new to the world of citizen deliberation, seems especially important for the challenges of deliberation and citizenship at a global level.

There were many research partners in the WWViews Alliance who had their particular interests and roles in contributing to the WWViews process. On the basis of the traditional conception of science as 'organized scepticism', tensions between practitioners and researchers can be expected, especially in areas where 'institutionalized attitudes' prevail (Merton, 1992, p278).[7] There was little evidence of such tensions in WWViews, however. Most of the academics involved in the project were from the fields of science studies or action research, where engagement in social interventions has been acknowledged as a viable research approach and scepticism towards non-deliberative forms of governance is commonly shared. However, it should be possible to find even more productive cooperation between practitioners and researchers in the future. The most evident areas of increased cooperation include: improving access between Western and non-Western partners; integrating evaluation as a core aspect of the project; coordinating and conducting the scientific analysis of the results; and contributing to a culturally sensitized design process. The continued involvement of social scientists can enhance the evolution of WWViews in important respects: increased reflexivity in design (methodological robustness), critical evaluation (scientific assessment and justification), scientific study (discursive perspectives and long-term reflexivity), and last but not least, human and financial resources.[8]

NGOs could be expected to be among the main stakeholders who seek to advance citizen participation in decision-making. In some cases NGOs indeed played a prominent role, as in Uruguay, where the NGO Simurg (dedicated to improved science–society interactions) took responsibility for organizing WWViews (Bortagaray et al). In many cases, however, NGOs seemed like 'value advocates'[9] who took the view that changes brought about by a WWViews type of citizen deliberation could clash with their values and interests. Three elements seem to be involved here. First, many advocacy organizations focus mainly on positions developed through extensive work and collaboration with other actors. Given this investment of time and in relationships, their inability to predict the outcome of WWViews naturally made some of them reluctant to partner with WWViews.[10] Second, a core WWViews philosophy was to offer a channel for citizens outside organized interests groups (such as industrial lobbyists or professional NGOs), and citizens or organizations that were considered too green were filtered out of the WWViews process. Third, NGOs had their own campaigns and 'citizen forums' that competed for the attention of the media and COP15 policy-makers, and that were the primary focus of their staff.[11]

Considering that many WWViews recommendations produced by average citizens called for stricter climate policies, the first NGO concern now seems less warranted. The second issue highlights the distinct roles that NGOs and the WWViews type of citizen participation aim to play, and the different channels used and postures adopted for informing policy-makers about the 'will of the people'. Finally, the unprecedented number of campaigns at COP15 and intense competition for attention towards citizens' voices on climate policy has led many of the contributors of this volume to propose new ways of cooperating, if not a 'new contract', between NGOs and WWViews types of forums.

There are many other actors – media, businesses, politicians, technology assessment units, think tanks, development organizations and funding agencies, to name prominent examples – that played a special role in the WWViews process and deserve closer attention. In the end, the most important actor in hearing the voice of citizens is the listener, the politicians who are responsible for devising policies and representing the values and interests of different sections of society.

Ideas for future development of global deliberations

The contributors to this volume make numerous proposals for developing future global deliberations based on the WWViews experience. In this section we will explore these practical implications in four key dimensions: organization and management, institutional cooperation, methodology and options for reconfiguring the deliberative elements (Figure 15.1). Recognizing that many of the proposals entail a risk of damaging the method that already passed its first test of vitality, our aim is to identify advantages and disadvantages of each proposed alternative. Before engaging these specifics, we will first discuss the relevance of a multilevel deliberation concept that is central in WWViews and many proposals in this volume.

Creating a multilevel approach to deliberation

Realizing that the universal model of WWViews responded imperfectly to the various needs of different localities, many of the contributors to this volume propose ways in which local concerns could be better linked to the global deliberation process. These point towards a multilevel deliberation approach, in which global issues and interests are more carefully configured with their local counterparts. As we will see, some issues are destined to remain global, while others can be addressed at multiple levels.

The proposal for including local issues on the global deliberation agenda (see Rask and Laihonen) aims at replacing – to varying degrees – the global agenda of deliberation with local-level issues and questions. The current model of WWViews did not

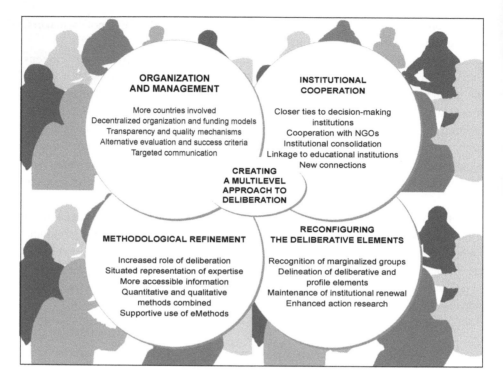

Figure 15.1 *Ideas for developing future transnational deliberations*

allow such localization except as an added component, as for example in the Italian WWViews site in Bologna, where participants met for an additional afternoon to discuss regional climate policy issues. The benefits of involving local issues include better (cognitive) access to the concrete questions related to the topic of deliberation, increased (social) motivation through the revelation of local interests, and enhanced (political) mobility through links to local dilemmas. The downside of this strategy is that it introduces (cognitive) asymmetries between countries and regions in the deliberation process, reducing the comparability of results, and it involves diffuse (social) concerns and (political) actions that may prove irrelevant at the global level of climate policy. Considering that a common transboundary problem is the basic reason for global deliberation, a significant replacement of global policy problems seems a remote idea, except for the Italian type of solution, where global and local questions are discussed somewhat separately.

Cooperation with both global and local policy actors and institutions in ways that help them become motivated and integrated into the deliberation process poses fewer contradictions. In the current WWViews model those levels were combined, for example, by targeting the policy recommendations to national climate negotiators. Another area with prospects of linking the global and local levels is the role of global citizens

Figure 15.2 *Organization and management*

that is created through the worldwide deliberation process. The emphasis in WWViews was on harmonized methods, questions and background materials, which all served the function of generating policy advice based on the (converging) views of the 'global citizenry'. This function, however, does not necessarily conflict with the idea of identifying particular issues that originate from different localities and introducing them on the global radar. The benefit in so doing lies in empowering the citizens by acknowledging not only their answers to specific questions, but their various concerns and perspectives as legitimate input to the global policy-making process.[12]

Developing the organization and management of global deliberations

The high costs of global deliberative processes speak for organizing them through decentralized networks, supported by some central resources and online knowledge-sharing venues. WWViews as a whole was not funded centrally, leaving each partner to find its own financing. Based on this experience, the distributed funding model seems likely to remain the most viable option for the future, especially if the processes become larger and more complicated (Andersson and Shahrokh).

There are two opposing managerial challenges that follow from this model. The first challenge is about controlling heterogeneity. In the context of heterogeneous backgrounds, motivations and expectations of the partners, it is important to ensure a high professional quality and integrity in the conduct of the deliberation processes. This calls for creating mechanisms of knowledge transfer, monitoring and partner feedback that help develop and maintain procedural validity. Ensuring high quality in the deliberation process at each site is the best insurance against a damaged reputation, and the key resource for political impact. Another aspect of the control challenges relates to the achievement of both scientific and political legitimacy amid a multitude of partners and audiences with different expectations. Visible forms of governance, such as an international steering committee, could be established to support scientific and political legitimacy. Such a committee might combine not only experts on issues as diverse as climate science (or scientific aspects of any other issue under consideration) and participatory policy processes, but a diverse sample of national partners and citizen participants from prior deliberations.

The second managerial challenge is about facilitating heterogeneity. The current model of WWViews was targeted at providing citizens across the globe an opportunity to influence global climate policy, and consequently, the function of deliberation was bound to that of policy advice. Since the partners of the WWViews Alliance, however, have diverse backgrounds and competences, they will most probably be more success-ful with some types of goals while encountering difficulties with others. For example, partners with a background in communication design are likely to be effective in raising awareness and reaching media; partners with experience in organizing interactive pro-cesses strong in enhancing social learning and interactions between different constituen-cies; partners operating in the field of policy advice clever in fostering policy pathways based on the results of deliberation (see Bechtold et al; Hennen et al, 2004, p63).

These diverse backgrounds could be recognized by acknowledging a plurality of project goals that the partners could prioritize according to their specific situations. This strategy could also help in finding synergies and complementary competences among the participating institutions. Also, partners could better manage and develop their practices with the development of criteria for evaluating success in a range of key dimensions, such as the effective inclusion of marginalized groups (Blue et al; Bal) or political impacts of participation in its various dimensions (Herriman et al). More generally, evaluation of pivotal project activities could become an integral aspect of future deliberations.[13]

Enhancing institutional cooperation

Effective institutional cooperation is the *sine qua non* of the vibrant network that the WWViews Alliance must become if it is to make a difference in global environmental governance. In the policy-making arena, additional venues and stronger ties are critical opportunities for institutional collaboration. UNFCCC is a central transnational body in global climate policy, but not necessarily the only place for deliberative governance.[14] Other alternatives could be more independent entities like the World Bank or World

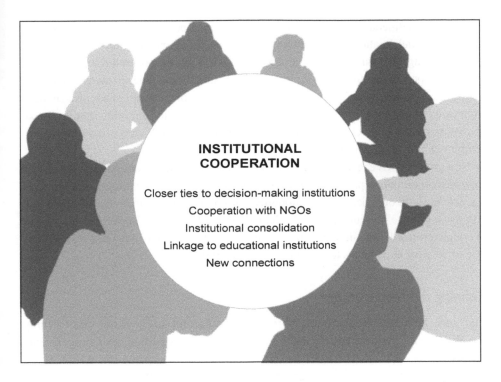

Figure 15.3 *Institutional cooperation*

Trade Organization that are less controlled by the states than UN bodies. Since positions brought to the global negotiations are developed at the national level, however, it is crucial to build stronger ties to national decision-makers. Among the challenges faced in doing this is timely delivery of the results. This requires effective planning for collaboration with policy-makers, such as an earlier start with pre-event briefings and advance commitments from decision-makers for post-event follow-ups (Herriman et al). To ensure that state functionaries actually use participatory results, individual champions willing to promote change in the culture of governance from within the state should be recruited into the network (Bal).

Cooperation with NGOs is another form of institutional collaboration recommended by the contributors. Organized interests are a legitimate part of any modern society (Agger et al), and civil society organizations can effectively raise issues and exercise 'soft transfer' of ideas that are underestimated by the political elite but still important for 'ordinary citizens' (see also Rask and Laihonen; Andersson and Shahrokh). Orchestrating the different ways of representing the interests of civil society, however, requires both conceptual and practical alignment of activities by NGOs and transnational deliberation bodies.

Institutional consolidation of global deliberation involves a shift from the world of one-off pilots to regular consultations. The positive experiences from at least six international deliberations add credibility to the idea of deliberative global governance (Andersson and Shahrokh). Expanding cooperation beyond institutions directed at policy-making opens a 'world of opportunity'. An important resource in this regard, already tapped in the first WWViews, is higher education. The growing interest in public engagement in this sector, and the in-kind resources and academic funding streams available to it, provide both motivation and capacity to collaborate in global deliberations. An instructive example is the integration of WWViews into a university seminar in Uruguay (Bortagaray et al).

New connections can also contribute to the visibility of transnational deliberation. A suggestive example is provided by the Boston (US) WWViews partner, where a local science museum organized a web-supported forum on global climate deliberation with a Parisian science museum and DBT (Schneider and Delborne). Impacts of new forms of cooperation are difficult to anticipate, but the examples given encourage the exploration of creative approaches to institutional collaboration.

Refining the methodology

The core elements in the contributors' analyses and suggestions for refining the WWViews method largely followed five alternative lines, and ranged from

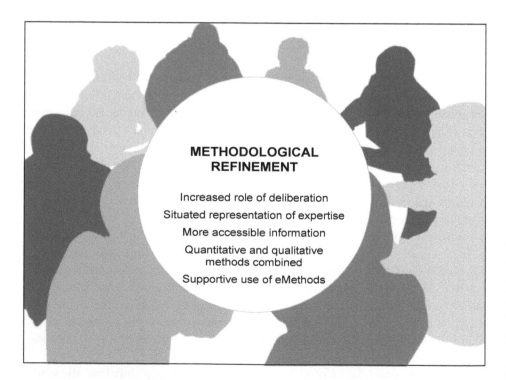

METHODOLOGICAL REFINEMENT

Increased role of deliberation
Situated representation of expertise
More accessible information
Quantitative and qualitative methods combined
Supportive use of eMethods

Figure 15.4 *Methodological refinement*

thoroughgoing changes to more focused refinement of particular elements. Given the myriad judgements involved in actually conducting global deliberations, many more could in principle be addressed (facilitation, analysis of results, etc.). Our focus on the most prominent issues is partly the product of limited space, but also reflects our sense that discussions of global deliberations could too easily become a narrowly methodological discourse.

The proposal to increase the role of deliberation is among the most thoroughgoing of the methodological recommendations (Goldschmidt et al). It basically suggests that the truly deliberative components of WWViews should be increased at the expense of 'decisionistic' voting. This proposal, resembling the European Citizens' Consultations model (see Andersson and Shahrokh), would make WWViews a fundamentally qualitative exercise yielding results and arguments that better explain why citizens prefer certain solutions. By emphasizing the role of argumentative or 'discursive representation' instead of statistical representation, the more deliberative model might enjoy greater methodological legitimacy. The downside of this proposal, in addition to the eclipse of the quantitative outcomes that politicians and the media want, is that increased deliberation (especially an additional international face-to-face event) would lead to increased procedural complexity and costs, as well as difficulties in aggregating and comparing the results across countries. Since the recommendation would entail a significant revision of the existing method, a wholesale adoption of the approach seems inconsistent with the evolutionary development of global deliberative governance that is generally supported in this volume. Nonetheless, the suggestion might inspire enhancements in the deliberative dimension of the citizen participation process.

Involving experts face-to-face with citizens is a more specific recommendation that aims to strengthen the scientific elements of the deliberation process, which relied on the information booklet and videos in WWViews. Engaging experts in the deliberation might help address the problems of a 'canonized' information booklet (Agger et al), as well as the uneven knowledge bases of participants noted by Goldschmidt et al, which is even more severe in the global South (Bal).

In our view, the introduction of experts in the local deliberation processes could help better communicate the scientific background of climate political debate and enhance reflection from multidisciplinary points of view. However, expert panels are difficult to manage and compose in a balanced way, and the risk is that they will just create new biases between countries and reduce comparability of the results. Instead of directly involving scientific experts in the deliberation process, it might make more sense to develop other strategies for strengthening scientifically informed dialogues. The call for more accessible and reflective information materials is one such strategy that focuses on the way in which scientific information is provided to the citizens. Bechtold et al propose that the current information materials could be made more accessible and balanced by using documentary films[15] and interviews as the primary channel for representing expertise, treating information booklets as a supportive resource. Being an incremental modification to the current WWViews concept, the recommendation should be comparatively easy to implement, with a low risk of methodological damage.

The recommendations for combining qualitative and quantitative methods in gauging citizens' views of public policy matters (Agger et al; Andersson and Shahrokh) emerge from the need to ensure demographic representation and methodological legitimacy of the WWViews method. Such proposals aim to enlarge the scope of citizens consulted through internet or other means without diminishing the centrality of face-to-face discussions. In our view the successful experiences from Deliberative Polling® and similar methods make consideration of such possibilities for future deliberations an important priority.[16] We can anticipate that methodological and organizational issues will arise in considering these innovations. If qualitative and quantitative techniques are used in a truly integrated manner, the complexity and costs of the process are likely to exceed the capacities of many current partners. If the quantitative consultation is used merely as a supplement to the qualitative deliberation, issues of the primacy, functionality and interpretation of two different consultations arise.

The recommendation to use internet-based communication and social networking technologies are directed at increasing the number of citizens that can be involved simultaneously in a participatory process. As Andersson and Shahrokh indicate, there are numerous pioneering e-engagement processes and supportive institutional frameworks, such as the Pan European eParticipation Network (PEP-NET), that could help develop online transnational deliberations. Even though this recommendation is primarily focused on a single component of the WWViews method, its effects can turn out to be radical rather than limited to the existing concept of deliberation. The added value of such tools seems considerable, so one key question is whether separate funding can be arranged to support their development and use. Another question is the pattern of inclusion and exclusion that specific e-engagement models generate.

Reconfiguring the deliberative elements

Some of the ideas and viewpoints discussed by the contributors provide alternative ways of understanding the rationale for deliberation and how it is linked to its social and policy contexts. The role of marginal groups, action research, institutional renewal and balancing between deliberation and profiling activities are four key dimensions that contributors have emphasized in new ways (Figure 15.6).

Each of these four dimensions has elements that are weighted towards external concerns in the social and policy environment as well as internal concerns among the network of practitioners. The relative weight put on these dimensions has much to do with whether the resulting recommendations envision improving the current model of WWViews with new components (gradual development), or constructing a completely new model (radical development). The emphasis in this volume has been on enhancements rather than wholesale revision.

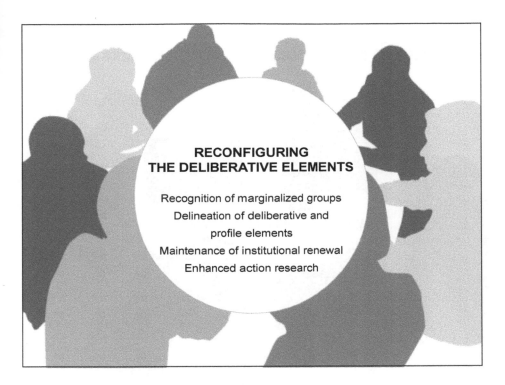

Figure 15.5 *Reconfiguring the deliberative elements*

Better recognition of socially marginalized groups is the focus of the 'empowerment model'. In their report on the Canadian WWViews, Blue et al describe how the random selection process in the Canadian WWViews was expanded through a targeted selection of northern and Aboriginal populations. Bal similarly argues that planning for deliberative recruitment in India needs to start with encouraging and supporting participation by women and people from lower castes. This empowerment aims to compensate for historically formed social inequalities through the deliberation process, and to avoid repeating or reinforcing existing power structures.

In its external dimension, the empowerment model encourages recruiting oppressed social groups and giving them more seats in the deliberation room than their respective share of the population would otherwise allow. Inside the deliberation process, the empowerment model emphasizes 'rituals of acknowledgement', for example, creating separate spaces for deliberation, recognizing different styles of argumentation, or recruiting women staff members, to support participation by the marginalized groups in the deliberation.

Considering the positive experience from the Canadian WWViews, the empowerment model provides important insights into how to support deliberations in culturally diverse contexts through nuanced facilitative ideas and gestures. Such actions seem

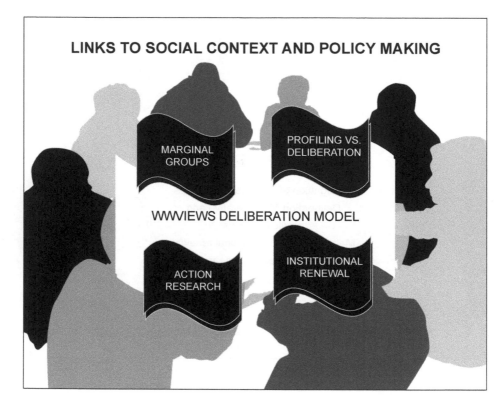

Figure 15.6 *Dimensions for reconfiguring the deliberation model*

more supportive than destructive to the current deliberation process, even where they involve seating participants from marginalized groups at separate tables. The number of seats allocated in this way, however, is not irrelevant. For example, assigning 10 per cent of the seats to oppressed groups that constitute only 3 or 4 per cent of the population could arguably improve 'discursive representation' and therefore strengthen the deliberation. Assigning 20 per cent of the seats in such a situation, on the other hand, would probably give the feel of a social movement rather than a citizen deliberation. Decisions on issues like these are likely to turn out best when they are made as informed judgements in a supportive community of practice.

A different but challenging front opens with the matter of delineating the deliberative and 'profile' elements of WWViews. By profiling elements we refer to activities that increase the visibility and media impact of the deliberation. One caution emerging from this volume is that strong focus on media attention and public relations efforts can become counter-productive to the deliberation process (Schneider and Delborne; cf Parkinson, 2005). The question of an appropriate balance between these elements, however, is complex.

Profiling in the external context of the deliberation means extensive media outreach activities, such as the identification of 'pegs' (themes designed to attract media interest) that are normally most effective when they align the story with prominent political conflicts and sensationalist narratives in the news. Integrating competencies in news media, advertising and public relations in the evolving WWViews Alliance would be the parallel internal development.

Considering the experience from many of the national WWViews sites, US (Schneider and Delborne) and Australia (Herriman et al) especially, the strategy of re-cruiting more people from the field of media relations seems well reasoned, since many of the practitioners report on their limited competences in this area. Seeking political influence primarily through media or overly emphasizing media visibility, on the other hand, seems a misplaced strategy for a deliberation exercise, since it undermines the premise that deliberations should provide fresh ideas outside partisan politics, not just throw fuel to the flames of an ongoing political debate.

Deliberation's role as a catalyst of institutional renewal also has its external and internal dimensions. The transformative potential of deliberation is illustrated by Bortagaray et al in their reporting of the new activities in the aftermath of the Uruguayan WWViews process. Not only were the curriculum of university education and thinking of the students of the Biology and Society course (in which context the local event was organized) affected, but other activities such as a new consensus conference on nuclear energy with national authorities and research organizations were organized. As Bortagaray et al report, 'Uruguay is taking its first steps toward active public participa-tion in S&T matters'.

Changing cultures of policy-making towards more open, participatory and demo-cratic modes is the aspiration behind most deliberation activities and philosophies. Given that many of the countries that participated in WWViews, China and Russia for example, have political systems that are hierarchical and authoritarian rather than participatory and democratic, it is interesting to see that deliberation processes are still found relevant. One explanation is that the 'transformative potential' of deliberation can be understood in rather different ways. To ensure that deliberation experiments remain attractive in the future, it is important to leave enough room for alternative political framings of such activities.

The internal aspect of institutional renewal relates to how the WWViews Alliance can ensure its own renewal and independence. The stabilization and institutionalization of a global deliberation process and network would support continuity, resources and access to institutions of decision-making. Such a development, on the other hand, may coincide with increased centralization, subordination and formalization of the deliberation concept, resulting in a tendency to turn meaningful participation into an encumbering ritual of bureaucracy. Considering the two risks – threat of institutional discontinuity versus loss of creative capacity – the former seems the real challenge in the current state of development.

The specific manner in which research is configured in the development of a social process such as global deliberation can have important consequences. Many members of the WWViews Alliance were 'action researchers' in the sense that they studied the deliberation process while also participating in its organization, even though this was not a part of the original project design.

While research is unlikely to provide correct answers to the many dynamic issues addressed in the course of conducting and interpreting a global deliberation, it can enhance organizational learning and interaction by providing a more informed base for decisions, enhancing transparency, and facilitating quicker adaptation. Future deliberations should therefore incorporate research efforts into the project from the outset. Internal resources for this include participants in the existing research consortium (many of whom are contributors to this volume), who are willing to develop evaluation (of the process) and analysis (of the results) in future WWViews. In the final section we discuss how research can contribute to the development of global deliberation.

Global deliberation: A mission for research[17]

An increasing number of the most severe environmental problems facing humanity are transboundary in character, making global deliberation a potentially critical tool for helping address the world's most severe problems. In our view, there are three different roles in which research can help in harnessing the potential of global deliberation. First, empirical research can generate an improved understanding of methods and instruments that are appropriate for organizing global deliberations on specific policy issues and problems. Second, it can enhance appreciation of the different expectations that prevail in diverse cultural settings regarding global deliberative governance. Finally, research can identify the institutional forces that limit the influence of deliberative actions and identify the space for transformation.

Empirical research[18] on the methodology of global deliberation

The methods for global (or transnational) deliberations can be empirically studied by analysing how WWViews was organized in different countries, or comparing the WWViews process to other transnational deliberative processes (see Andersson and Shahrokh). WWViews aimed at creating a voice for 'global citizens'; the following questions represent a range of approaches to understanding how such a voice can be built:

- How do different recruitment methods influence the selection of participants and the deliberation?
- How can global citizenship be constructed through global deliberation?
- What is the role of global deliberations in promoting action to mitigate and adapt to climate change?

Empirical research on transnational deliberations still lags behind theoretical treatments of the subject, and recognizing that many chapters in this volume are focused on national perspectives, there is considerable room for enquiries into the transnational dimension of the deliberation method.

Social research on culturally constructed expectations of global deliberation

The transformation of political systems is a slow process and based on culturally constructed expectations of how political institutions work. The expectations that different actors have in supporting such a transformation is a key to improved understanding of the 'meaning' of deliberative democracy and a critical tool in fitting these processes to their cultural contexts. WWViews was unique in its global scope and inclusion of developing countries. The expectations prevailing in the different contexts can be studied through following these types of questions:

- What are the motivations and experiences of those partners who joined WWViews from the global South, including those who were unable to generate the resources to conduct the deliberation?
- What criteria should be applied to evaluate the success of global deliberation in different political cultures?
- How should different audiences for deliberative results be identified and approached?
- What can we learn from practices in different cultures?

The common denominator of these questions is that they can best be answered through an intersubjective evaluation of the roles and possibilities of deliberative action. A continued discussion with relevant actors and stakeholders can help answer these questions, ideally with implications for policy relevance.

The institutional contexts of global deliberation

Public deliberation can potentially become an integral part of global politics, for example as an indicator for timely no-regrets policies by anticipating the future direction of public opinion and facilitating early responses to cumulative problems. Understanding that the methodological functions and social expectations of WWViews-type activity is one side of the coin, the other is the need to reflect on the institutional forces that limit or support the transformation towards more deliberative global governance. The following questions can help in reflection on these institutional conditions:

- What relations between deliberation and policy-making would be optimal in different scenarios/contexts/topical discourses?

- What are the decision processes that global deliberations aim to influence, and what are the policy pathways that lead to such impact?
- How should governance of the World Wide Views Alliance evolve?

Realizing the potential of global deliberation requires not only continued research efforts but also calls for self-reflection by political actors on how WWViews-type activity fits into the extant institutional landscape, and what is required to make it fit there.

The contributions to this volume by a group of international practitioners who facilitated the WWViews project and scholars of environmental governance provide a unique account of the first citizen participation in global environmental governance. We feel privileged in having the chance to document and reflect these experiences, and we express our gratitude to all the practitioners, academics and citizens involved in WWViews. Understanding the value of this first global deliberation was identified as the goal of this volume in the introductory chapter. We can now add that the value of the deliberation in 2009 and especially the potential significance of future WWViews are important for deliberative global governance. We will be delighted if this volume plays a part in adding to that potential.

Notes

1 In many countries several organizations collaborated to conduct a single deliberation. Of the 38 countries where deliberations were held, half were in the global South.
2 Undated author names refer to chapters in this volume.
3 Similar progressions of deliberative practices from sub-national to national levels could be mapped outside Europe as well (see e.g. the account for Australia by Herriman et al), but there is no equivalent to the Europe-wide deliberations in other global regions.
4 For example, Todashi Kobayashi, 'WWViews: Western-centered citizen participation?', Program Chair's Plenary, Society for Social Studies of Science Annual Meeting, Tokyo, 25 August 2010.
5 Interviews by S. Cozzens and R. Worthington at US White House Office of Science and Technology Policy and White House Council on Environmental Quality, March 2010.
6 In explaining their protection of the term 'Citizens Jury' as a registered trademark, the Jefferson Center refers to their aspiration to maintain a high level of integrity and trust-worthiness in the deliberation process; see www.jefferson-center.org/index.asp?Type=B_BASIC&SEC={2BD10C3C-90AF-438C-B04F-88682B6393BE}. The 'Deliberative Polling' trademark is registered by James S. Fishkin. Any fees derived from it are used to support research at the Center for Deliberative Democracy (Gastil and Levine, 2005, pv).
7 Science, according to Merton has the methodological and institutional mandate of temporarily suspended judgements and detached scrutiny of beliefs – leading occasionally to conflicts with other institutions (Merton, 1992, p277).
8 It is clear that researchers from different backgrounds exhibit different attitudes towards campaigns such as WWViews. In the context of designing social interventions, it might be

helpful to distinguish between 'inside' and 'outside' experts. The inside experts have adopted the role of exercising critical reflection while taking part in social interventions, whereas outside experts focus on providing criticism and reflections based on more detached theoretical perspectives.

9   For the definitions of 'stakeholders' and 'value advocates', see Renn (1992) and von Wartburg and Liew (1999).

10 A case in point is a Green Party member of the Finnish Parliament who feared that the WWViews results might be less radical than his party's stance on the issues.

11 The founder of the highly visible 350.org campaign, Bill McKibben, was a WWViews Ambassador and spoke at the WWViews side event at COP15. Nonetheless, 350.org never publicized the WWViews results on its website, despite a commitment to do so. This probably was due to the enormous logistical pressures faced by 350.org, which organized the largest day of citizen action in history on 24 October 2009, less than a month after the WWViews results were available.

12 The 488 citizens' recommendations in WWViews provide a unique perspective on the local perspectives, indicating also an interesting tendency of people to seek power from consumerism when lacking power in global political questions (Lammi et al); as many of the contributors argue, however, the role of those more deliberative products were downplayed in the current process.

13 This practice is already evolving in the US, e.g. Archon Fung et al, 'Public impacts: Evaluating the outcomes of the CaliforniaSpeaks statewide conversation on health care reform', at http://californiaspeaks.org/wp-content/_data/n_0002/resources/live/CaSpks%20 Evaluation%20Report.pdf, accessed 2 December 2010.

14 This argument was presented by John Dryzek at a WWViews workshop in Snekkersten, Denmark; see http://blogit.kuluttajatutkimus.fi/ilmastoareena/?page_id=558; for further discussions, see Dryzek et al, 2011.

15 At the Snekkersten workshop the idea of partnering with companies such as National Geographic or the BBC was raised: http://blogit.kuluttajatutkimus.fi/ilmastoareena/?page_id=558.

16 These possibilities did receive consideration in developing WWViews, but never gained traction.

17 This section builds largely on the discussions among WWViews researchers at a workshop in Snekkersten, Denmark 2010; see http://blogit.kuluttajatutkimus.fi/ ilmastoareena/?page_id=558.

18 The WWViews project (and related research-based documentation) generated extensive empirical data consisting of project manager interviews, exit surveys, pre-surveys, recordings of the table-level deliberations, media follow-up documents, etc. Access to these materials can be requested from the contributors to this volume.

References

Aron, R. (1995) 'The anarchical order of power', *Daedalus*, vol 124, no 3, pp27–52

Bexel, M., Tallberg, J. and Uhlin, A. (2010) 'Democracy in global governance: The promises and pitfalls of transnational actors', *Global Governance*, vol 16, no 1, pp81–101

Boltanski, L. and Chiapello, E. (2005) *The New Spirit of Capitalism*, Verso, London

DBT (2010a) *World Wide Views Method*, Danish Board of Technology, www.wwviews.org/node/10, accessed 12 December 2010

DBT (2010b) *World Wide Views Design*, Danish Board of Technology, www.wwviews.org/node/248, accessed 12 December 2010

Dryzek, J. S. (2006) *Deliberative Global Politics: Discourse and Democracy in a Divided World*, Polity Press, Cambridge

Dryzek, J. S., Bächtiger, A. and Milewicz, K. (2011) 'Toward a deliberative global citizens assembly', *Global Policy*, vol 2, no 1, pp33–42

Gastil, J. and Levine, P. (eds) (2005) *The Deliberative Democracy Handbook: Strategies for Effective Civic Engagement in the 21st Century*, Jossey-Bass, San Francisco.

Hennen, L., Bellucci, S., Brloznik, R., Cope, D., Cruz-Castro, L., Karapiperis, T., Ladikas, M., Klüver, L., Sans-Menéndez, L., Staman, J., Stephan, S. and Szapiro, T. (2004) 'Toward a framework for assessing the impact of technology assessment', in M. Decker and M. Ladikas (eds) *Bridges Between Science, Society and Policy: Technology Assessment – Methods and Impacts*, Springer, Berlin

Klüver, L., Bellucci, S., Berloznik, R., Bütschi, D., Carius, R., Cope, D., Decker, M., Gram, S., Grunwald, A., Hennen, L., Karapiperas, T., Ladikas, M., Machleidt, P., Sans-Menéndez, L., Peeters, W., Staman, J., Stephan, S., Szapiro, T., Stayaert, S. and van Est, R. (2004) 'Technology assessment in Europe: Conclusions and wider perspectives', in M. Decker and M. Ladikas (eds) *Bridges Between Science, Society and Policy: Technology Assessment – Methods and Impacts*, Springer, Berlin

Merton, R. K. (1992) *The Sociology of Science: Theoretical and Empirical Investigations*, The University of Chicago Press, Chicago and London

Parkinson, J. (2005) 'Rickety bridges: Using the media in deliberative democracy', *British Journal of Political Science*, vol 36, pp175–183

Renn, O. (1992) 'The social arena concept of risk debates', in S. Krimsky and D. Golding (eds) *Social Theories of Risk*, Praeger, Westport, CT

Sclove, R. (2010) *Reinventing Technology Assessment: A 21st Century Model*, Woodrow Wilson International Center for Scholars, Washington, DC, www.wilsoncenter.org/index.cfm?topic_id=1414&fuseaction=topics.documents&group_id=271875

von Wartburg, W. P. and Liew, J. (1999) *Gene Technology and Social Acceptance*, University Press of America, Lanham, MD

Young, I. M. (1989) 'Polity and group difference: A critique of the ideal of universal citizenship', *Ethics*, vol 99, no 2, pp250–274

Appendix

Comparison of
International Deliberations
(Andersson and Shahrokh, Chapter 4)

Meeting of Minds

Year	*2006*
Website	www.meetingmindseurope.eu
Summary	The Meeting of Minds process was one of the first deliberative processes that spanned more than one European country. It pioneered techniques in cross-boundary engagement and gave citizens a chance to engage on the controversial topic of brain science.
Issue focus	Brain science
Background	Funding was provided in 2003 to run a process which aimed to involve European citizens in publicly discussing the issue of brain science with relevant research experts, stakeholders and representatives of European decision-making organizations.
Rationale	Advances in brain sciences and the controversies they generate are gaining increased public attention, but policy-makers have not yet engaged these issues significantly. The initiative aimed to give relevant inputs into the wider public debate on brain science. The process focused on developing participatory technology assessment, the envisaged effects being: to enhance social learning; to stimulate public debate; and to provide policy advice (Meeting of Minds website).
Funder	European Commission, Directorate General for Communication (DG Comm) as well as a consortium of foundations across Europe.
Delivery bodies	King Baudouin Foundation (Brussels) and partners including NGOs and universities.
Participants	126 in total, selected by national project teams in each country.

Countries   Nine European Union member states.

Methodology   In each participating country three national events were held as well as two
                    joint European meetings (in 2005 and early 2006). The citizens explored
                    the issue of brain science, generating a shared framework and setting out
                    those aspects of brain science that need to be examined further. National
                    panels took these proposals home and continued working on them to produce
                    conclusions on the desirability of brain science and put forward issues for the
                    European agenda. The second European meeting took on board the national
                    conclusions and recommendations and ran further with them, producing
                    a European assessment report on brain research issues. The participants
                    discussed areas of overlap, agreements and disagreements, the underlying
                    reasons for them and what could be learned. The methodology aimed to
                    integrate national assessments into the central European process, making the
                    national events a vital component in the international deliberation (Meeting
                    of Minds website).

Results   The results of the discussions were incorporated into a European report
                    with conclusions and recommendations that was handed over to high-level
                    European officials and representatives of the European scientific and re-
                    search community. One of the aims of this exercise was to create an ongoing
                    dialogue at European level between the general public and policy-makers
                    on science-related matters in order to stimulate public debate and induce
                    self-reflection (Meeting of Minds website).

                    Anecdotal evidence gathered orally by the authors indicates that despite
                    being the oldest of the processes and the one which involved the fewest
                    participants, Meeting of Minds is more frequently quoted by subject experts
                    than many other of the projects in this study. The choice of discussing an
                    emerging issue meant that it was particularly difficult to interest policy-
                    makers and the media. Only a small number of stakeholder groups were
                    actively engaged with the theme, making it more difficult to develop firm
                    policy recommendations (Meeting of Minds website).

European Citizens' Panels

Year   *2006–2007*
Website   www.citizenspanel.eu

Summary   This pilot was a bottom-up initiative that engaged 337 citizens from ten
                    different regions of Europe to deliberate on policies that affect rural areas.

Issue focus   The research question was: 'What roles are there for rural areas in tomor-
                    row's Europe?' Specific issues included agricultural policy, climate change
                    and social welfare policy, for example access to housing.

Background   EU agricultural and rural policy consumes a large part of the Union's budget while problems in rural areas persist, raising both budgetary and social concerns.

Rationale   1 EU enlargement involves countries where these rural issues are even more salient than among current members.
                    2 Shared democratic identity is weak in the EU and further challenged by enlargement.
                    3 Organizers wanted to design and test a method for European-level citizen participation.

Funder   European Commission (Directorate General for Education and Culture), EU Committee of the Regions, central or local government agencies in each participating country, plus foundations in the UK, Ireland and The Netherlands.

Delivery bodies   Consortium of universities, NGOs and think tanks coordinated by the European Association for Information on Local Development (a Belgian not-for-profit organization).

Participants   337 randomly selected citizens took part in eight regional panels, 87 of whom subsequently met in Brussels to prepare a joint statement.

Countries   Nine European countries (ten regions).

Methodology   Events were structured around regions, some of which crossed national borders, as in the case of the Carpathians (Hungary/Slovakia). The eight regional panel events all made use of a common methodology. The pan-European session identified key common themes from the regional deliberations in small group and plenary sessions over three days, generating 24 recommendations.

Results   The regional panels drew together recommendation reports with the help of professional facilitators and these were presented to appropriate regional decision-makers. The European recommendations were presented in a report to key European decision-makers, as well as institutions that supported the panel's work at regional level. The same report was disseminated to the wider public.

Tomorrow's Europe: Deliberative Poll®

Year   *2007*
Website   www.tomorrowseurope.eu

Summary   This project is the first time that Deliberative Polling® has been used in a cross-boundary setting. Unlike the other processes outlined here, where separate national meetings were held, Tomorrow's Europe involved bringing participants together in one location.

Issue focus	'Key social and foreign policy issues affecting the future of the EU and its member states'.
Background	The Deliberative Poll® was commissioned under the European Commission's 'Plan D' (Democracy, Dialogue and Debate) programme. Plan D's impetus is largely the product of French and Dutch voters' rejection of the European Constitution in June 2005. This prompted the heads of state and government to call for a 'period of reflection' in order to enable member states to initiate national debates on the future of Europe. Plan D sought to enable this 'reflection' by encouraging such national debates.
Rationale	1 Ahead of the 2009 European Parliamentary elections the Deliberative Poll® was to demonstrate that there is a clear, effective line of communication between EU policy-makers and citizens across borders. 2 The process was to test the methodology for the first time at a cross-national level.
Funder	European Commission (DG Comm) and a number of foundations, private companies and government bodies across Europe.
Delivery bodies	Initiated and coordinated by Notre Europe (Paris-based think tank) with a wide range of partners, including Stanford University (US) and think tanks from across Europe.
Participants	A poll of 3500 demographically representative citizens were surveyed on their views of EU policies and 362 of these citizens were brought together for a face-to-face conference in Brussels. Because of the different population sizes some countries were only represented by two or three individuals.
Countries	Participants from 27 countries, all brought to one event in one country.
Methodology	The process began with an initial poll on key matters for the future of the EU, in particular Europe's economic and social dimension, involving a sample of 3500 citizens. The participants had received balanced information on the issues in advance. Small-group and plenary discussions involved simultaneous translation in 23 languages. The participants developed questions which were put to experts and political representatives. Parts of the weekend events were collected for broadcast on television in taped and edited form. A final poll then measured citizens' opinions after the deliberations.
Results	Tomorrow's Europe demonstrated that it is possible to bring participants from a wide and diverse range of countries together in one room and incorporate important face-to-face cross-cultural interchanges. The process involved the German EU presidency, the European Parliament and the European and Social Committee and also had extensive media coverage including: *Le Monde*, *The Guardian*, *Der Spiegel* and Reuters. The use of statistically robust samples was appreciated by some policy-makers, although the disparity between large and small countries (for example 47 German participants versus 4 Slovenians) means that the validity only holds true at the EU level and not at the national level.

European Citizens' Consultations (ECC) I and II

Year	*2007 and 2009*
Website	www.european-citizens-consultations.eu/files/ECC%2009%20Project%20 Description%20final.pdf
Summary	These two processes were the first European Union (EU)-wide deliberative engagement processes which involved meetings in all EU countries. They allowed randomly selected members of the public to make recommendations for the future of the EU in various areas. They also represent the first example of a cross-national deliberation process to have been repeated.
Issue focus	First process (2007): energy and environment, family and social welfare and the EU's global role and immigration. Second process (2009): 'What can the EU do to shape our economic and social future in a globalized world?'
Background	These projects were part of the European Commission's 'Plan D' (Democracy, Dialogue, and Debate). See information in Tomorrow's Europe above.
Rationale	The European Citizens' Consultations were seen as a way of reconnecting the EU to citizens across the Union. It was also seen as a pilot to develop a new methodology to help the EU to build a wide consensus on its future policies.
Funder	European Commission (DG Comm) as well as a consortium of foundations across Europe.
Delivery bodies	King Baudouin Foundation and partner organizations in the 27 EU countries. Partners included universities, think tanks, NGOs and private-sector companies.
Participants	Depending on size of country 30, 50 or 100 randomly selected participants took part. Around 1500 citizens took part in each ECC process in total. In the 2009 process the recruitment was standardized and coordinated across all countries.
Countries	All 27 EU countries had ECC events in 2007 and 2009. The national meetings took place over three weekends followed by a central Brussels-based event (in 2007 one citizen per country took part in the Brussels event, in 2009 10 per cent of participants from each country took part).
Methodology	The European Citizens' Consultations coordinated discussions across national borders and shared the results of voting between events. The methodology incorporated elements of the 21st Century Town Hall Meeting and world cafés (see www.peopleandparticipation.net/display/Methods/World+Cafe). Participants were divided into facilitated groups. Priorities and suggestions were fed into laptops and voted on. Some events made use of electronic voting technology. The 27 national consultations were synthesized into a single report taking into account the diversity of different member states at

a European-level event, held in Brussels. The ECC marked a massive step change from the Meeting of Minds by involving participants from all 27 member states and by increasing the numbers involved substantially.

In 2009 the process incorporated an online forum open to anyone to complement the face-to-face consultations with extended deliberation outside the research event itself. Online debate and voting preceded and influenced the deliberative event. This allowed a much larger group of people to take part in a more limited way.

Results The two ECC processes proved that ambitious cross-national processes are possible to plan and run, and that cultural differences do not prohibit comparable deliberation. The ECC can be seen as a pioneering model for dialogue processes spanning the EU which dealt with far more difficult logistical challenges than the preceding processes.

The 2007 project made an impact on communication policy within the European Union, as communication on Debate Europe emphasized that the European Commission supports future 'pan-European participatory democracy projects holding citizens' consultations in each Member State and establishing a common set of conclusions at the European Level'. The process was re-commissioned by the European Commission, whereas all other processes studied have been one-off pilots. Both ECC programmes involved several heads of state, ministers, European commissioners, members of European and national parliaments, plus observers and volunteers in all member states. ECC 2009 aimed to bring longevity to the citizen involvement created by the process by making it more structured within the European decision-making system. The second instalment added new innovations such as an online debate; an internal online forum for participants in the national deliberations and additional regional outreach in order to enhance the dissemination of the results (Project Outline: European Citizens' Consultations). Continued debate was held with MEPs, Commission representatives and foundations. This resulted in the Commission supporting the addition of citizens' consultations to the consultative toolbox of the EU – next to opinion polls and stakeholder consultations. However, funding is a barrier.

WWViews on Global Warming (WWViews)

Year   *2009*
Website   www.wwviews.org

Summary   The first truly global deliberative engagement process covering multiple continents. Allowed roughly 4000 members of the public in 38 countries spanning six continents to make recommendations for the United Nations Climate Change Conference (COP15) discussions in December 2009.

Issue focus   Climate change (with a focus on the COP15 Conference in Copenhagen).

Background	The COP15 Conference aimed to develop a successor to the Kyoto Protocol. It was one of the largest global policy processes ever. WWViews was set up to provide a forum for citizens' views to be considered.
Rationale	Enabling the citizens' voices that are missing from the global policy process to influence international decision making.
Funder	The Danish and the Norwegian foreign ministries provided some central funding for coordination but, unlike the other projects that we have studied, there was no central funding for events to draw on. Each in-country delivery partner had to find funding for their own event.
Delivery bodies	Consortium led by the Danish Board of Technology. Partners included academic bodies, government bodies, foundations and NGOs.
Participants	About 4000 participants spread over 44 events (the target number for each event was 100). Participants were randomly selected to meet demographic criteria for each country. A few different options were given for recruitment to adjust to differing contexts. The 100 citizens at each meeting reflected a demographic distribution of their country; they were also screened to exclude climate change experts, scientists and lobbyists.
Countries	38 from six continents.
Methodology	One event per location; all held on 26 September 2009. Prior to the event and on the day all participants received the same set of balanced information. All meetings were structured according to the same format in order to ensure comparability. Discussions took place at tables of five to eight people and were led by facilitators. Participants voted on 12 questions under four themes, making international quantitative comparison possible. An online tool allowed the outcome of the meetings to be posted to the WWViews web page immediately. As a result there was instant accessibility.
Results	The WWViews process was working at an unprecedented global scale. Despite the diversity of contexts and partner organizations, most events went according to plan. The WWViews target groups were politicians, negotiators and interest groups at COP15. The dissemination of the results was performed by all national and regional WWViews partners who each made their own plans for reaching these groups. A number of the global partners succeeded in building support for the process through country-specific policy briefings and meetings with their country's delegates in order to present their reports. The process was held less than three months before the main UN conference. By this point much of the negotiating had already happened and feeding citizens' voices into the process was much more challenging. The dilemma is that if WWViews had been held too early it would have been much more difficult to engage citizens effectively in the issues. However, the closer to the main event the process got, the more the policy positions of national governments and their negotiating teams were firmed up.

Index

Milton Keynes UK
Ingram Content Group UK Ltd.
UKHW031144141024
449569UK00024B/1094

9 781849 713795